T0297415

Terrorism, Security, and Computation

Series Editor
V.S. Subrahmanian

More information about this series at http://www.springer.com/series/11955

V.S. Subrahmanian • Michael Ovelgönne
Tudor Dumitras • B. Aditya Prakash

The Global
Cyber-Vulnerability Report

 Springer

V.S. Subrahmanian
University of Maryland
College Park, MD, USA

Michael Ovelgönne
University of Maryland
College Park, MD, USA

Tudor Dumitras
University of Maryland
College Park, MD, USA

B. Aditya Prakash
Department of Computer Science
Virginia Tech
Blacksburg, VA, USA

ISSN 2197-8778   ISSN 2197-8786   (electronic)
Terrorism, Security, and Computation
ISBN 978-3-319-25758-7   ISBN 978-3-319-25760-0   (eBook)
DOI 10.1007/978-3-319-25760-0

Library of Congress Control Number: 2015954657

Springer Cham Heidelberg New York Dordrecht London

Printed on acid-free paper

Springer International Publishing AG Switzerland is part of Springer Science+Business Media (www.springer.com)

Foreword

This book does something that almost no one has ever done before (and certainly not to this extent): it seriously, and empirically, tests some common assumptions about cybersecurity. It succeeds in measuring, in objective terms, the degree of cybersecurity vulnerabilities of different states and compares them scientifically.

The public awareness of cybersecurity is growing very fast. Still, people, even experts, have gross misconceptions about the relative vulnerability of certain countries. The authors of this book succeed in empirically refuting many of those wrong beliefs. They do it by thorough analysis of huge amount of data—more than 20 billion telemetry reports monitoring over four million host machines for 2 years—provided by Symantec. Then they focus on 44 countries and measure the actual number of attacks per machine and percentage of attacked machines in a given country.

Surprisingly enough, many common beliefs turn to be false. The most vulnerable countries are not those usually mentioned in the media but are India, South Korea, Saudi Arabia, China, Malaysia, and Russia (in that order). This well-established result is a consequence of what I like to call the *Paradox of Public Awareness*: the more secured a country is, the more vulnerable it is conceived to be by the public. That is because high level of cybersecurity implies high rate of malware detection and, therefore, greater *image* of vulnerability.

The authors do not stop at providing a methodology for using hard data to precisely measure cyber-vulnerabilities. They carry out the analysis and use the results to prove that there is a negative correlation between the wealth of a country (measured by per capita GDP, Human Development Index, level of health care, etc.) and its level of vulnerability: the more rich and developed a country is, the less vulnerable it is.

Perhaps one of the most interesting parts of this book is the discussion of national policies suggested by the book's findings. The authors touch areas like education (going down even to elementary schools), the need for *Cybersecurity Authority* at the highest national level, etc. *Human behavior* turns to be highly important and it corroborates my own main lesson learned after 25 years of experience in this field— the *Interdisciplinary Principle of Cybersecurity*—although solutions to cybersecurity problems are usually technological, the problems are never purely technological.

One should take into account human behavior (as individuals and as societies), legal problems, psychology, and more.

Let me finish the foreword of this highly valuable book with a comment on the future. No one can predict what it will look like, but we know that the role of computers will continue to be substantial in the future. Some of us have visions of Internet of Things (IOT), smart homes and smart cities, autonomous cars, intelligent robots, etc. Nothing of these wonderful visions will occur without a sufficient level of cybersecurity. Therefore, *cybersecurity is not a standalone discipline designed to secure our machines. It should be viewed as an enabler to our future.* Like the Moon, computerization has a bright side as well as a dark side. While the bright side is expressed by those visions of IOT, robots, etc., the dark side is expressed by cybersecurity, and we better remember what Pink Floyd said about it: "There is no dark side of the moon, really. Matter of fact, it's all dark."

Tel-Aviv University, Tel-Aviv, Israel   Isaac Ben-Israel
Chairman, Israeli Space Agency, Tel Aviv-Yafo, Israel
Founder (former), National Cyber Bureau
in the Prime Minister's Office, Jerusalem, Israel

Preface

It is becoming increasingly clear that cybersecurity will become one of the dominant security risks of the twenty-first century. It would not be an overstatement to say that virtually every person in the world who has access to any kind of computing device will be at risk and will likely experience adverse cyber-events. With the ubiquity of mobile phones around the world, as well as the expected explosion of computing capabilities through the advent of the "Internet of Things," we expect cyber-risk to grow as hackers find ever more creative ways to wreak havoc in the lives of unsuspecting users.

This book is about *risk*. Despite the heavy investment in cybersecurity by countries around the world, few of these countries have solid quantitative data assessing their risk and relating it to the general risk profile around the world. For instance, the deputy director general of an EU nation's intelligence agency recently attended a talk by the first author and remarked afterward that he did not have access to the type of data that we had when writing this book.

Our goal in writing this book was to take the first steps to quantify the risk to different countries via some simple, easy to understand, data-driven metrics. Thanks to Symantec's Worldwide Intelligence Network Environment (WINE), we had access to an amazing dataset that included over 20 billion telemetry records (both binary reputation and malware reports) from over four million hosts per year over a 2-year period. Symantec collected this dataset from users who had "opted in" to this collection using their antivirus products such as Norton Antivirus. We note that the data analyzed in this book is available for follow-up research as the reference dataset WINE-2013-001.

We note that this book represents just an *initial step* toward quantifying cyber-vulnerability of nations. We only studied Windows hosts and consumer machines. Some countries may have advanced military capabilities and/or have robust businesses that are less at risk than the ordinary consumers located in those countries. Moreover, no mobile hosts were studied—a flaw that we hope to correct in coming years. Last but not least, we note that all monitored hosts had Symantec antivirus products such as Norton Antivirus running on them. Attacks on these hosts were detected and blocked by Symantec's antivirus products. It is possible that other hosts in the countries we studied that did not have an antivirus product running on them had a different cybersecurity profile. To guard against this, we only studied countries with at least 500 hosts reporting per year, so that the sample size was large enough. In fact, for most countries, we had well over 1000 hosts reporting—and our 2011 dataset had over 1000 hosts monitored in each of the 44 countries we studied. We believe the results of our study will provide important quantitative, data-driven insights to policy makers in at least these 44 countries.

We conclude with a note of thanks to all of those who helped us in our research. The research in Chap. 4 was funded in part by Cliff Wang and the Army Research Office under ARO Grant Number W911NF1410358, by Sukarno Mertoguno of the Office of Naval Research under ONR Grant Number N00014-15-1-2007, and the Maryland Procurement Office under contract number H98230-14-C-0137.

A big vote of thanks is due to Matt Elder of Symantec Corporation who spent an extensive amount of time reviewing the content of this book and providing valuable feedback. Without Matt's patient explanation of some of the Symantec WINE data and his hard work, this book would not have been possible.

At the University of Maryland, we would like to thank PhD student Noseong Park for helping in producing some of the figures and Barbara Lewis for helping in formatting this document. Aaron Mannes proofread the document diligently. At Virginia Tech., we would like to thank MS student Benjamin Wang for helping in some of the research presented in Chap. 4. At Springer, we would like to thank Susan Lagerstrom-Fife and Jennifer Malat for their kind support.

College Park, MD   V.S. Subrahmanian
College Park, MD   Michael Ovelgönne
College Park, MD   Tudor Dumitras
Blacksburg, VA   B. Aditya Prakash
August 30, 2015

Contents

Chapter 1
Introduction

In a keynote address at the IFIP Security conference in Marrakech in May 2014, the first author asked the audience to name the five countries that they believed were the most vulnerable nations on earth from the perspective of cyber-security for ordinary consumers. Not governments. Not enterprises. Everyday consumers, the man on the street or you and me.

The answers were surprising. We heard, in alphabetical order, France, Germany, India, Japan, the Netherlands, the UK, and the USA. Many of these answers reflected the countries of origin of the respondents. A similar thing happened at the kickoff of a new multi-million dollar cybersecurity grant at the University of Maryland. And with the exception of India, all respondents were wrong. Worse still, except for India, all of the countries they listed are amongst the most cyber-secure in the world for individual consumers.

The same trend continued when the first author asked the same question at the IEEE Joint Intelligent and Security Informatics Conferences (JISIC) in The Hague in September 2014.

Gross misconceptions about cyber-vulnerability are widespread, even, as can be seen from these brief episodes, amongst the world's foremost cybersecurity experts.

This book provides a methodology to answer the following questions.

1. *Can we use hard data to clinically and precisely quantify the cyber-vulnerability of nations*? Yes. We leverage 2 years of data from Symantec consisting of over 20 billion telemetry records (both binary reputation and malware reports) and over four million (per year) users worldwide to precisely quantify the cyber-vulnerability of 44 countries that include all the major OECD countries, as well as the major developing economies. Though we had access to data that involved many other countries, we eliminated countries from our study for which we deemed the data insufficient. Chapter 5 provides statistics and a brief discussion of cyber-attacks on each of these 44 countries.

2. *Can we assess the types of malware that target different countries*? For each of the 44 countries listed above, we were able to precisely quantify the nature of the

© Springer International Publishing Switzerland 2015
V.S. Subrahmanian et al., *The Global Cyber-Vulnerability Report*, Terrorism, Security, and Computation, DOI 10.1007/978-3-319-25760-0_1

malware that targets those countries. We classified malware into six types: worms, Trojans, misleading software (e.g. fake anti-virus software), spyware, adware, and a catch-all "other" category. We precisely assess the nature of the threat faced by each country. Chapter 5 assesses the types of malware targeting each of the 44 countries in our study.

3. *Is the wealth of a nation measured in terms of per-capita GDP related to cyber-vulnerability of those nations?* Here, the answer is a clear and unambiguous yes. Rich countries generally fare better, both in terms of the percentage of hosts attacked and in terms of the number of malware attacks per host. In comparison, poor countries are more susceptible to attacks.

4. *Are certain classes of consumers more vulnerable to cyber-attacks than others?* The answer is yes. We study four classes of human users—software developers, professionals (e.g. doctors, lawyers, businessmen), gamers, and "others". A fifth group includes "all" users. It is important to note that these classes can have overlapping members—for instance, a person may be both a gamer and a professional. Our results show that software developers are the most vulnerable, with gamers coming in second.

5. *Is the development status of a nation measured by the UNDP's Human Development Index (HDI) related to the cyber-vulnerability of a country?* The answer is a clear and unambiguous "yes". The HDI measures the development status of a country via three broad criteria: the health of individuals in the country and the country's ability to measure health care, the knowledge of the people in the country as measured by various educational opportunities, and the economy of the country. We are able to establish definitively that the greater a country's HDI, the less cyber-vulnerable the country is.

6. *Is software piracy related to or a cause of increased cyber-vulnerability of nations?* The answer here is much less clear. Though software piracy is certainly correlated with an increased number of malware attacks on host machines when considered alone (as also found in [1]), when considered in conjunction with other contributing factors, its impact becomes much less. Separating the impact of piracy from other GDP-related factors is complex. When we take into account the fact that most attacks are web-based (which we know from other studies), piracy appears to have a low overall influence.

7. *What national policies do these findings suggest? How should countries better protect their ordinary citizens from cyber-vulnerability?* Our study suggests that the cyber-vulnerability of countries as a whole is positively correlated to the number of downloaded binaries and negatively correlated with per capita GDP. However, in addition to these factors, they are also positively correlated to the number of binaries present on hosts, the number of rare binaries present on hosts, and the number of unsigned binaries. Simply put, practicing good cyber-hygiene should be taught in elementary schools from the age of 6 onwards. Governments and businesses should make a 1-hour cyber-hygiene video that explains best practices for their employees and encourage them to share these habits with their family, friends, and professional colleagues.

Those who wish to get a high-level overview of the results of this book only need to read this one chapter which summarizes the main findings of the book without going into the technical details. Those interested in results about a specific nation should read the appropriate section of Chap. 5.

The rest of this book is organized as follows.

1. Chapter 2 provides informal definitions of basic cybersecurity concepts. It explains the intuitive meaning of different types of malware and how host machines can be infected with malware in a variety of ways. It is intended to be accessible to an informed and intelligent policy maker, not a cybersecurity expert.
2. Chapter 3 describes the data and methodology we used in great detail and also provides the statistical evidence required to substantiate the results described in this chapter.
3. Chapter 4 describes human behaviors that are related to cyber-vulnerability. The results suggest that downloading large numbers of binaries (either intentionally or unintentionally) exposes users to increased risk, as do connections to unfamiliar networks, and installing rare binaries (software that is not frequently used by others).
4. Chapter 5 assesses the cyber-vulnerability of each of 44 countries. This chapter constitutes the main contribution of the book. We only considered countries in which we had data on at least 500 host machines. For most countries, the number of host machines studied ran into the tens of thousands.

Before proceeding further, we emphasize three things.

1. First, correlation is not causation. The results linking various countries' situations (e.g. GDP, software piracy, human development index, educational levels) reflect correlations, not causation. A study involving the establishment of causal relationships between these variables and the cyber-vulnerability of these nations would require the ability to perform a controlled experiment in which several country related variables like GDP, software piracy rates, human development index, educational levels, etc. were varied. Unfortunately there is no feasible way for us (or even much more powerful bodies like the World Bank or UNDP or ITU) to carry out such experiments.
2. Second, we note that this book focuses on the consumer side of cybersecurity.
3. Third, the measures of cyber-vulnerability that we consider in this book are based on file-based attacks. In particular, network-based attacks detected by the host IPS are not included.

We do not assess the risk to countries' military and intelligence establishments, nor do we assess the cyber-vulnerability of businesses and enterprises in these countries.

1.1   Data and Methodology

Our study used Symantec's Worldwide Intelligence Network Environment (WINE) data set (version WINE-2013-001) collected by Symantec's anti-virus products such as Norton Anti-Virus. Individuals who buy such products are asked if they would be willing, on an opt-in basis, to allow Symantec to collect various kinds of data about their machine and behavior in order to improve detection using aggregated telemetry. Symantec uses rigorous privacy protocols to ensure that they protect each individual's privacy. As a consequence, every Norton Anti-Virus user who has "opted in" provides valuable feedback to Symantec. The WINE data is a subset of the Norton Anti-Virus data that has been significantly scrubbed by Symantec in order to remove Personally Identifying Information. In addition, to further protect the data, Symantec requires that all studies by outside experts (including us) both be approved by Symantec and be conducted at a Symantec facility. Any results generated by the study must be approved for export outside the facility by Symantec. No raw data ever leaves Symantec.

The WINE data that we used in our study consists of two parts:

1. WINE's binary reputation data set includes information on all binary executables, both benign and malicious, that were downloaded or otherwise installed on a machine. The data includes information such as the time-stamp when the file was created, the country in which the host resides, an MD5 or SHA2 hash of the binary, and the URL from which the binary was downloaded (if applicable).
2. WINE's anti-virus telemetry data set includes reports about host-based threats (e.g. viruses, worms, Trojans) detected by Symantec's anti-virus products. It classifies a piece of malware into a threat type (Trojan, worm, misleading software, spyware, ad-ware and other), includes MD5 and/or SHA2 hashes, and the technique used to detect the malware (e.g. signature scanning vs. behavioral analysis).

We note that the data analyzed in this book is available for follow-on research as the reference data set WINE-2013-001.

Because most Anti-Virus providers including Symantec are interested in having very low false positive rates (i.e. when they identify a binary to be malware, they want to be very sure that it is), binaries labeled as being malicious are highly likely to be malicious—this was confirmed verbally by Symantec personnel. This will turn out to be very important for our study.

Our study used a total of 2 years of data from September 2009 to August 2011. In the September 2009 to August 2010 period (which, for the sake of simplicity, we refer to as the "2010" data), we studied a total of 4.23M host machines, while in the September 2010 to August 2011 period (which for the sake of simplicity, we call the "2011" data), we studied 4.14M machines in total. Any host machine for which we had less than 100 days of data was discarded. Our study considers every country for which at least 500 machines were monitored. However, as the table below shows, for most countries, the number of machines considered was significantly larger than that. In fact, in 2011, at least 1000 hosts were monitored in all 44 countries in our study.

	2010	2011
Number of countries with 500–1000 hosts monitored	6	0
Number of countries with 1000–10,000 hosts monitored	27	17
Number of countries with over 10K–100K hosts monitored	8	23
Number of countries with over 100K hosts monitored	3	4

Throughout this book, we use the terms "attack" and "infection attempt" interchangeably. An attack on a host represents a detection by Norton Anti-Virus of a piece of malware on that host—so all incidents represented as attacks or infection attempts are attacks that have been detected by Norton Anti-Virus and prevented from infecting the host. In cases of known malware, this detection represents a blocked attack or infection attempt. We note that in many cases, malware is reported days, weeks, months, or even years after an initial infection attempt occurs. For instance, [2] reports that 18 previously undiscovered pieces of zero-day malware existed in the wild (i.e. were undetected) for 312 days before they are detected. Machines which are not protected by an anti-virus product are likely to be infected by such attacks for significant periods of time.

The WINE data was augmented with open source data from:

- The World Bank's Data Catalog which provides (among other things) worldwide economic data;
- The UNDP's Human Development Report which provides extensive data on the health care, educational status, and economic status on people in virtually every country in the world;
- The Business Software Alliance which reports on software piracy.

1.2   The World's Most Cyber-Vulnerable Countries

In a book that is largely about bad news, we start with some good news. Half (50.2% to be exact) of the hosts that we studied were free of any malware attacks. That is indeed good news. However, this number is distorted by the large number of US hosts that we studied. When we look at individual countries, the percentage of hosts where no malware attacks were detected ranged for 10–65%. Even this may be better news than many may have hoped for. Though many measures can be used to assess the cyber-vulnerability of countries, we used two basic measures.

1. *Average number of infection attempts per host.* One measure of the cyber-vulnerability of a country is the average number of pieces of malware detected by the anti-virus software on a host machine.
2. *Percentage of machines attacked.* A second measure of the cyber-vulnerability of a country is the number of host machines on which infection attempts were detected.

Figure 1.1 shows a pictorial depiction of the cyber-vulnerability of the world in terms of number of cyber-attacks per host machine. *We emphasize again that*

2010

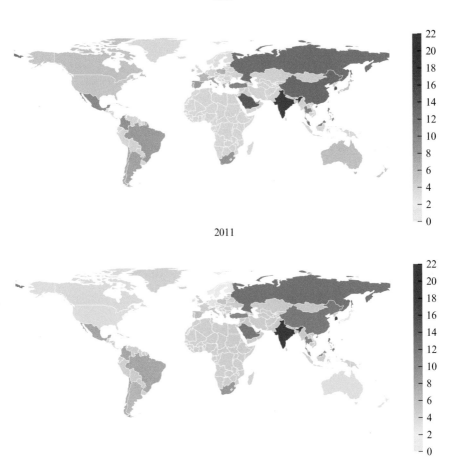

2011

Fig. 1.1 Attack frequencies per host for the whole world for 2010 and 2011

throughout this book, we use the terms "attack" and "infection attempt" inter-changeably to refer to a piece of malware targeted at a machine. Because Norton Anti-Virus is deployed on the machines in our study, these figures represent attacks that were eventually detected and nullified by Norton Anti-Virus. An unprotected machine in a country may well have been infected by such attacks.

We see from Fig. 1.1 that the cyber-vulnerability of most countries decreased when measured in terms of the number of attacks per host machine. Australia, Canada, China, Japan, Russia and the US all show marked improvements from 2010 to 2011 though other countries like Brazil and Argentina experienced increases. For European readers, Fig. 1.2 shows a map of Europe color-coded to show average number of attacks per host machine for each of the 2 years, 2010 and 2011. Here too, there is room for optimism as the average number of attacks per host machine in countries like Portugal, Spain, France, and Germany all decreased during this time period.

Fig. 1.2 Attack
frequencies per host in
Europe

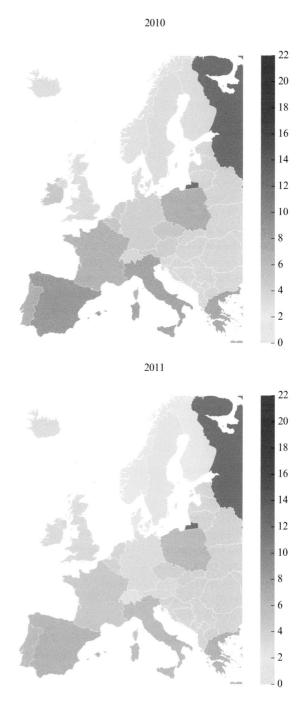

Which countries were the most cyber-vulnerable during this period? Table 1.1 below shows the ten most cyber-vulnerable countries according to both the measures listed above.

Table 1.1 may be a bit confusing at first because our two measures of vulnerability show variations in the years 2010 and 2011. Let us look at the ranks of all the countries involved according to these two measures for both years. Table 1.2 shows us this situation.

According to these rankings, India, South Korea, China, and Saudi Arabia stand out. Irrespective of the measure used or the years considered, these countries are consistently in the top seven according to every ranking and every year. If we drop Saudi Arabia, we see that India, China, and South Korea are among the top five most vulnerable nations according to both measures—and in both years. Comparing 2010 and 2011, S. Korea saw a huge increase in the average number of attacks per host as well as an increase in percentage of hosts that were attacked. According to a Symantec report [3] released in June 2013, there were 4 years of attacks (including our study period) orchestrated by North Korea on South Korea via a group called DarkSeoul. India actually saw a very small drop in the second measure, though the number of attacks per host did go up. China improved—both the number of attackers per host and the percentage of attacked hosts went down.

Table 1.3 shows the best ten countries from a cyber-vulnerability perspective: the top ranks are all dominated by Scandinavian countries with Switzerland and Germany not far behind. Northern European countries dominate—and furthermore, less economically developed nations like the Czech Republic do very well.

The reduction in Japan's cyber-vulnerability from 2010 to 2011 seen in Table 1.3 is notable. The average number of attacks on hosts in Japan dropped by over 50% in just 1 year, while the percentage of attacked machines dropped by almost 25%, significant improvements over very short periods of time.

Table 1.4 below ranks the top ten countries from a cybersecurity perspective. Irrespective of the measure of vulnerability used, Norway, Denmark and Finland are consistently in the top three. Japan also does well when it is ranked in terms of the percentage of attacked hosts, reaching rank 3 in 2010 and rank 1 in 2011. However, machines that were attacked in Japan tend to be populated with a larger number of malware binaries than the Scandinavian countries. The US comes in consistently around 10–11, irrespective of the ranking method used. In fact, the US, Netherlands, Australia and the Czech Republic all have similar rankings suggesting that the US is doing well, but has room for improvement.

In summary, looking across both 2010 and 2011, and examining both measures of cyber-vulnerability:

- **the most cyber-vulnerable nations** are India, S. Korea, Saudi Arabia, China, Malaysia and Russia. India and S. Korea are the most cyber-vulnerable with slight variations between 2010 and 2011, and Malaysia and Russia compete to decide who makes it into the top five with slight variations between the year considered and the type of cyber-vulnerability measure used.
- **the least cyber-vulnerable nations** are Norway, Denmark, Finland, Sweden and Switzerland with Norway and Denmark competing for the top spot and with Germany coming in sixth, just outside the top five.

Table 1.1 Ten most cyber-vulnerable countries

Country	2010 Avg Atks Per host	2010 % Hosts Attacked	2011 Avg Atks Per host	2011 % Hosts Attacked
India	17.93	0.827	19.53	0.809
S. Korea	15.24	0.819	21.63	0.913
Saudi Arabia	14.77	0.807	12.17	0.761
China	12.92	0.875	11.63	0.825
Malaysia	12.56	0.763	9.68	0.719
Russia	12.5	0.805	12.50	0.794
Turkey	10.7	0.775	10.82	0.752
Thailand	10.57	0.698	11.01	0.695
Philippines	10	0.729	12.98	0.764
Mexico	9.88	0.746	7.96	0.688
Colombia	9.38	0.771	7.85	0.73
Brazil	8.27	0.77	7.82	0.767
Taiwan	9.4	0.751	7.85	0.689

Table 1.2 Ranking of the ten most cyber-vulnerable countries

Country	2010 Avg Atks Per host	2010 % Hosts Attacked	2011 Avg Atks Per host	2011 % Hosts Attacked
India	1	2	2	3
S. Korea	2	3	1	1
Saudi Arabia	3	4	5	7
China	4	1	5	2
Malaysia	5	9	9	10
Russia	6	5	4	4
Turkey	7	6	8	8
Thailand	8	13	8	11
Philippines	9	12	3	6
Mexico	10	12	10	13

1.3   The Relationship Between Cyber-Attacks and GDP

Figure 1.3 shows a chart of the relationship between per capita GDP of countries and the number of cyber-attacks per host for 2010. A similar chart shows the situation in 2011.

We see an obvious inverse relationship between per capita GDP and the number of cyber-attacks per host. Simply put, and contrary to the feeling amongst many that citizens of rich countries are more heavily targeted, it turns out that citizens of countries with low GDPs are more vulnerable to cyber-attacks than citizens of rich countries.

Table 1.3 Ten least cyber-vulnerable countries

	2010	2010	2011	2011
	Avg Atks	% Hosts	Avg Atks	% Hosts
Country	Per host	Attacked	Per host	Attacked
Norway	2.87	0.437	2.20	0.364
Denmark	3.07	0.4	2.28	0.358
Finland	3.13	0.415	2.11	0.328
Sweden	3.17	0.463	2.41	0.378
Switzerland	3.35	0.435	2.41	0.387
Germany	4.09	0.517	3.09	0.435
Austria	4.17	0.526	3.17	0.468
Netherlands	4.36	0.523	3.70	4.89
Czech Republic	4.62	0.574	3.94	0.515
USA	4.86	0.555	3.51	0.485
Japan	5.84	0.417	2.64	3.20
New Zealand	5.07	5.26	2.97	4.46
Australia	5.69	0.577	3.38	0.491

Table 1.4 Ranking of the ten least cyber-vulnerable countries

	2010	2010	2011	2011
	Avg Atks	% Hosts	Avg Atks	% Hosts
Country	Per host	Attacked	Per host	Attacked
Norway	1	5	2	4
Denmark	2	1	3	2
Finland	3	2	1	2
Sweden	4	6	3–4	6
Switzerland	5	4	3–4	6
Germany	6	7	8	7
Austria	7	8–9	9	9
Netherlands	8	8	12	11
Czech Republic	9	12	12	13
USA	10	11	11	10
Japan	11	3	6	1
New Zealand	12	8–9	7	8
Australia	13	13	10	12

The same observation also holds when we consider the fraction of attacked hosts (Fig. 1.4).

Thus, our results show high correlation, both via a univariate statistical analysis and a multi-variate statistical analysis, that per capita GDP is inversely correlated with both the number of pieces of malware attacks on a per host basis, and the percentage of attacked hosts in a country. National GDP is also inversely correlated with the number of infection attempts on a host (Pearson correlation coefficient of −0.74 in 2010 and −0.69 in 2011).

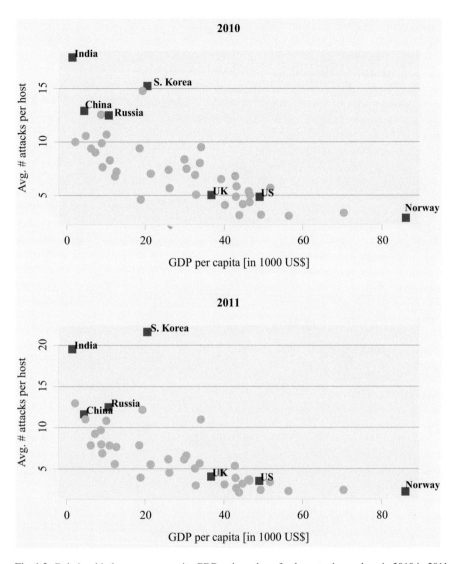

Fig. 1.3 Relationship between per capita GDP and number of cyber-attacks per host in 2010 in 2011

1.4   The Relationship between Cyber-Attacks and the Human Development Index

The Human Development Index (HDI) for short is a composite measure developed by the United Nations Development Programme (or UNDP). HDI measures the development of a country by examining a large number of variables related to public health, education, and economic factors. High HDIs generally imply well developed economies. Figure 1.5 shows the relationship between HDI and the average number

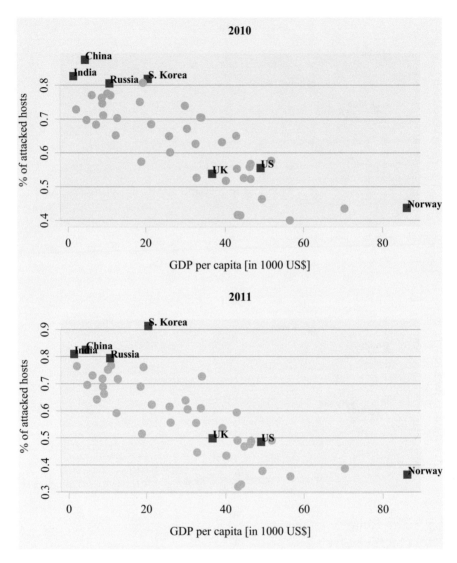

Fig. 1.4 Relationship between per capita GDP and percentage of attacked hosts in 2010 and 2011

of attacks per machine, while Fig. 1.6 shows HDI measured against the percentage of attacked hosts in a country.

Both figures show us the trend indicated already by GDP. Countries with a high HDI tend to have a low number of attacks per host (Pearson correlation coefficient: −0.69 in both 2010 and 2011) and a low percentage of attacked hosts (Pearson correlation coefficient: −0.71 in both years).

We note that HDI is clearly linked to GDP. Countries that have a high GDP can afford to spend more on their citizens' health care and their citizens' education. Thus, this indicator is closely linked to GDP and each influences the other.

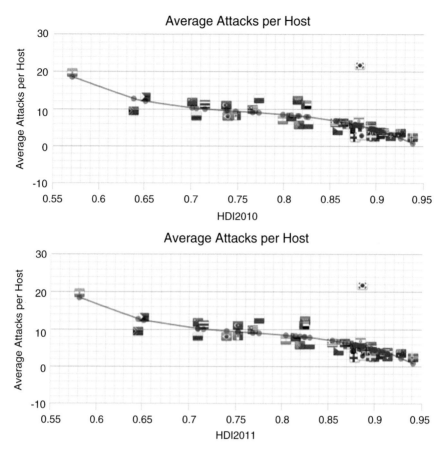

Fig. 1.5 Relationship between per HDI and number of cyber-attacks per host in 2010 and 2011

1.5   The Relationship between Cyber-Attacks and HDI Health Index

The HDI health index is a measure of the life expectancy of individuals on a country by country basis. Figures 1.7 and 1.8 shows the relationship between the health index of countries and both the number of cyber-attacks per host and the number of attacked hosts in 2010 and 2011. The HDI health index of a country measures the ability of the country to deliver services to its people. If a country cannot protect its citizens from disease, one may be justified doubting its ability to protect its citizens from cyber-attacks.

The situation in this case is a bit more nuanced than in the case of the HDI as a whole. Again an inverse correlation between the HI of countries and both the percentage of attacked hosts (Pearson correlation coefficient: −0.53 in 2010 and

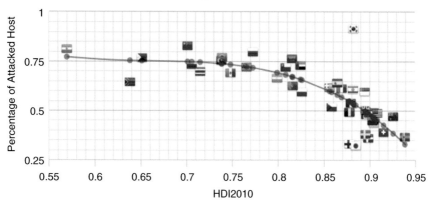

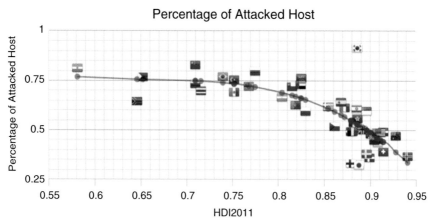

Fig. 1.6 Relationship between HDI and percentage of attacked hosts in 2010 and 2011

−0.54 in 2011) and the average number of attacks per host (Pearson correlation coefficient: −0.56 in 2010 and −0.57 in 2011). Though this may sound like a weaker negative correlation than in the case of the GDP, these figures are slightly skewed by the South Africa's extremely low health index. If we ignore South Africa, the country on the extreme left of Fig. 1.7, we see a clear inverse correlation between HI and the number of cyber-attacks per host. Countries with a better HI (likely wealthier countries) have lower number of cyber-attacks per host. A similar comment applies with respect to Fig. 1.8.

Interestingly, South Africa has a much higher per capita GDP (about 5 times that of India in 2011) but during the same year, its HDI health index (0.531) was significantly lower than that of India (0.702). This suggests that India did a better job providing health services to its citizens than S. Africa despite having a much lower GDP. Thus, even though there are obvious links between per capita GDP and the

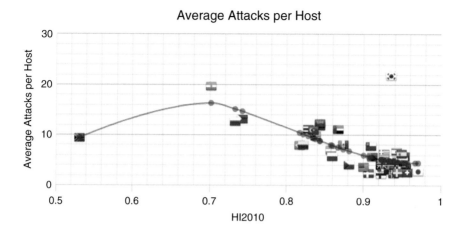

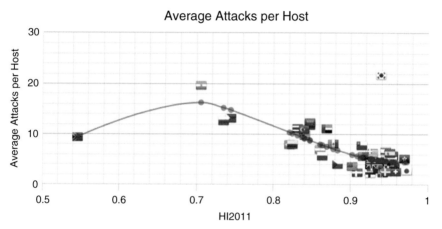

Fig. 1.7 Relationship between Health Index (HI) and number of cyber-attacks per host in 2010 and 2011

HDI Health Index, the HDI Health Index may say something about a country's government's ability to deliver basic services.

1.6   The Relationship Between Cyber-Attacks and Education

We examined a number of education related measures in the UNDP Human Development reports and studied their relationship with cyber-vulnerability. We note that these education measures are also closely linked to GDP and the Human Development Index of countries.

  We consider each of these in the subsections below.

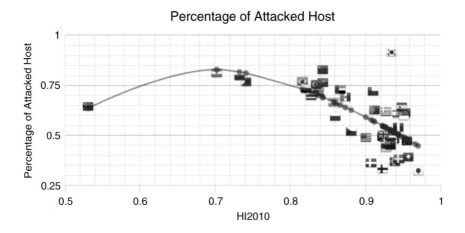

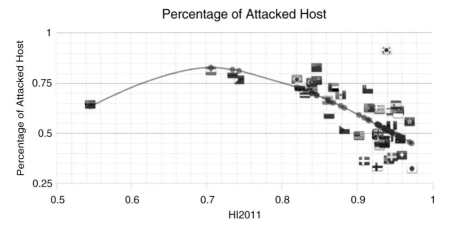

Fig. 1.8 Relationship between Health Index (HI) and percentage of attacked hosts in 2010 and 2011

1.6.1   The Overall Education Index

The Education Index (EI) component of the HDI combines the mean years of schooling and the expected number of years of schooling that individuals in a country might have (Figs. 1.9 and 1.10).

Figures 1.11 and 1.12 shows the relationship between health index of countries and both the number of cyber-attacks per host and the number of attacked hosts in 2010 and 2011.

In both of these figures, we observe an inverse correlation between the HDI Education Index of countries and both the percentage of attacked hosts (Pearson correlation coefficient: −0.63 in 2010 and −0.61 in 2011) and the average number of attacks per host (Pearson correlation coefficient: −0.64 in 2010 and −0.61 in 2011).

Interestingly, when we return to the case of India vs. South Africa discussed earlier, we see that S. Africa actually has a higher HDI Education Index than India, suggesting that it is doing a better job than India in educating its population.

1.6.2 Combined Gross Enrollment in Education

We also looked at the relationship between cyber-attacks and the combined gross enrollment in education (CGEE). The CGEE is reported in UNDP's Human Development Report as a percentage of the population at the appropriate age that are actually enrolled in either primary, secondary, or tertiary educational institutions, regardless of age and gender.

Figures 1.11 and 1.12 shows the relationship between CGEE of countries and both the number of cyber-attacks per host and the number of attacked hosts in 2010 and 2011.

In both of these figures, we observe an inverse correlation between the HDI Combined Gross Enrolment in Education (CGEE) of countries and both the percentage of attacked hosts (Pearson correlation coefficient: −0.50 in 2010 and −0.46 in 2011) and the average number of attacks per host (Pearson correlation coefficient: −0.48 in 2010 and −0.42 in 2011). This suggests that CGEE may not be as predictive of cyber-vulnerability than some other factors.

1.6.3 Expected Years of Schooling

The HDI Expected Years of Schooling (EYS) measure is exactly as it sounds. It measures the number of years that a child entering primary school can expect to receive if enrollment patterns do not change during the child's lifetime.

Figures 1.13 and 1.14 shows the relationship between CGEE of countries and both the number of cyber-attacks per host and the number of attacked hosts in 2010 and 2011.

We see here that as the expected years of schooling goes up, the average number of attacks per host declines steadily (Pearson correlation coefficient: 0.62 in 2010 and 0.60 in 2011). The case is similar with the percentage of attacked hosts except that for two countries, Australia and New Zealand, which buck the trend somewhat (Pearson correlation coefficient: 0.62 in 2010 and 0.60 in 2011).

1.7 Is Software Piracy Relevant to Cyber-Risk?

Several large organizations (e.g. Microsoft and the FBI) have asserted a link between the level of software piracy in countries and the cyber-vulnerability of those countries [4]. For instance, the FBI in a consumer alert warns "the American people

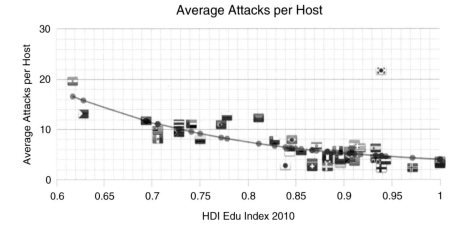

Fig. 1.9 Relationship between per HDI Education Index (EI) and number of cyber-attacks per host in 2010 and 2011

about the real possibility that illegally copied software, including counterfeit products made to look authentic, could contain malware." [5].

Our study shows that these assertions are correct on the surface—a simple univariate analysis of the link between software piracy rates as reported by the Business Software Alliance [6] and cyber-vulnerability shows that increased piracy rates are positively correlated to an increased number of both malware per host and an increased fraction of attacks per host in the country (Pearson correlation coefficient of 0.7 in 2010 and 0.65 in 2011). This is shown in Figs. 1.15 and 1.16 and is also consistent with past work [1] reporting piracy as a "cause" of increased vulnerability.

However, as we will see later in Chap. 3, a multi-variate regression analysis looking at number of cyber-attacks per host shows that software piracy is an insignificant factor in the regression, suggesting that software piracy is more a function of per-capita GDP (inversely correlated) than a cause of cyber-vulnerability.

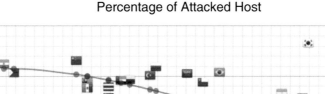

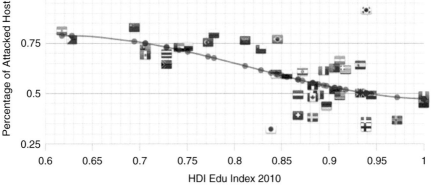

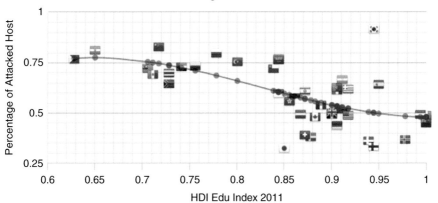

Fig. 1.10 Relationship between HDI Education Index (EI) and percentage of attacked hosts in 2010 and 2011

1.8 Some Related Work

Though the WINE data that Symantec provided us access to is very rich, the authors rely on Symantec to provide a representative sample of their overall data. As the third author was working for Symantec when WINE was created, we believe that this is a reasonable assumption. Nonetheless, we wondered whether malware identification methods may vary from one cybersecurity vendor to another. For instance, McAfee has a Global Threat Intelligence Network that they use to gather data that appears to have some similarities with WINE. The McAfee Threats Report is issued every quarter and we went through a sample of these reports.

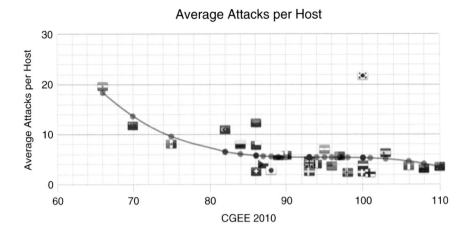

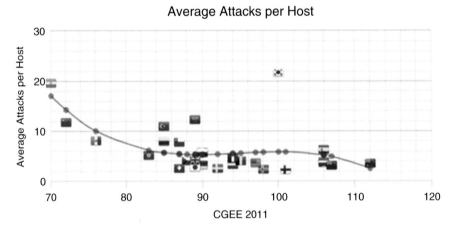

Fig. 1.11 Relationship between per HDI Combined Gross Education Enrolment (CGEE) and number of cyber-attacks per host in 2010 and 2011

For instance, in the McAfee Threats Report issued in the first quarter of 2013 [7], they noted that they had about 128M samples in their malware database (aggregated over several years)—but in the first quarter of 2013 alone, they saw 14M new types of malware. This number went up to over 30M new types of malware detected in each of the first two quarters of 2014 [8, p. 20] and by the end of the second quarter, McAfee's malware collection exceeded 250M samples.

McAfee's report asserts that "the United States is both the source and the target of much of the Internet's malicious activity." [7, p. 13]. They assert that the United States hosts the largest number of botnet servers, suspect URLs, phishing sites, spam URLs, and PDF exploits, but that the US is also the largest victim of SQL-injection attacks. There are some differences in these ratings from other reports—for instance, Symantec's April 2014 threat report [9, p. 93] says that the fraction of

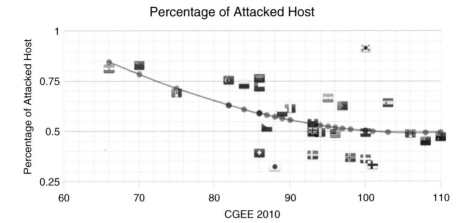

Fig. 1.12 Relationship between HDI Combined Gross Education Enrolment (CGEE) and percentage of attacked hosts in 2010 and 2011

email traffic identified as phishing related is highest for emails originating in South Africa, the UK, and Italy, respectively. Though these reports appear to be different from ours on the surface, in reality they are consistent with our report because we normalize attack numbers within a country by number of hosts (i.e. we report percentage of machines that are attacked and average number of attacks per machine) while they do not.

The McAfee 2014 third quarter threat report states that "North America continues to host more suspect content than any other region" [8, p. 25]. A similar report from Symantec [10, p. 46] asserts that "the US and China have the most densely populated bot populations", providing further insight into the origin of the attacks — their report lists Italy, Taiwan, and Brazil as rounding out the top five in terms of having the highest bot populations.

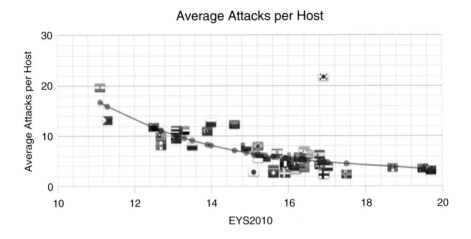

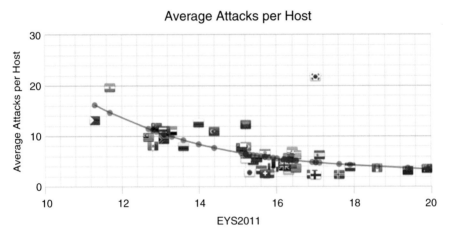

Fig. 1.13 Relationship between per HDI Expected Years on Schooling (EYS) and number of cyber-attacks per host in 2010 and 2011

These important industry reports provide valuable insight into the attacker side of the equation, examining where attacks are coming from. Unfortunately, we did not have access to the raw data that these companies used to generate the excellent graphics and summaries that they generated in these reports, especially those that deal with identifying the origins of attacks.

1.9   National Cybersecurity Policies: A Quick Summary

Cybersecurity is a global problem, one that transcends international boundaries. Cyber-criminals in Romania may attack individuals in the US, while pedophiles in the US may try to find victims in poorer countries. Because of the international,

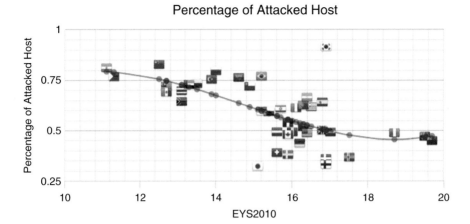

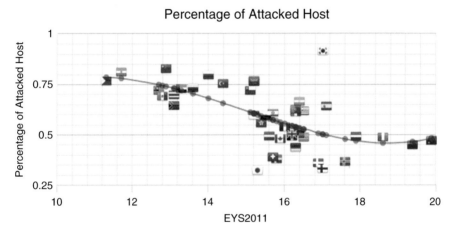

Fig. 1.14 Relationship between HDI Expected Years of Schooling (EYS) and percentage of attacked hosts in 2010 and 2011

cross-border nature of cyber-crime, nations have a responsibility to provide the best possible cybersecurity environment for their citizens.

Chapter 5 describes the cybersecurity policy of each of 44 countries. These include most of the large developed economies of the world (OECD countries) and most of the major developing economies including major emerging economies. The following factors are common across these 44 countries:

a. Almost all of these 44 countries have a well-articulated national cybersecurity policy—however, there are some exceptions.
b. There are variations in how countries have organized government response to secure cyber-space. In some countries (e.g. Israel), there is a centralized authority, while in other countries, responsibility to secure cyber-space is distributed across multiple ministries. Nonetheless, even in countries where responsibility

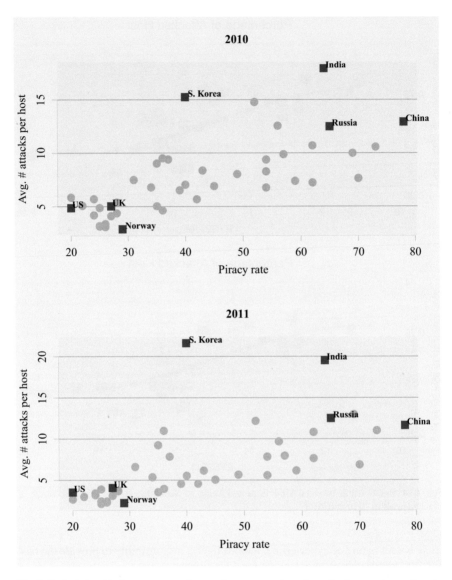

Fig. 1.15 Relationship between per software piracy rate and number of cyber-attacks per host in 2010 and 2011

for securing cyber-space spans multiple ministries, there are higher level governance structures in place to coordinate on a cross-ministry or cross-institutional basis. Nonetheless, such distributed structures run the risk that large-scale cyber-attacks will fall through the cracks, especially as cyber-attacks happen in very short time frames.

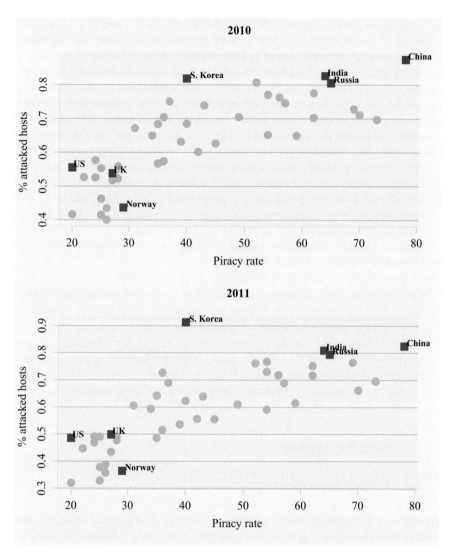

Fig. 1.16 Relationship between per software piracy rate and number of cyber-attacks per host in 2010 and 2011

c. There are significant variations in the governance structures involved in cybersecurity policy-making. In some countries such as the US, a national coordinator for cybersecurity sits in the White House and the head of US Cyber-Command is also the head of the National Security Agency, unifying intelligence and military functions. Some countries have no cyber-czar and instead have committees or panels that report to Prime Ministers. Other countries have a designated Minister who is responsible for cybersecurity, typically as part of a portfolio that involves other functions. Some countries put cyber-defense within their military command structure.

d. Virtually all countries recognize that cybersecurity is a cross-border issue and recognize the importance of cybersecurity partnerships. However, the operationalization of cybersecurity issues is a major challenge in cross-border efforts.
e. Virtually all countries recognize that cybersecurity requires a public-private partnership (PPP) involving businesses and governments. However, the role of universities in such PPPs is not well articulated in most national cybersecurity policies. Countries have been compromised by nation-state attacks that channeled their attacks through universities (e.g. the recent Regin attack on India). In addition, few countries offer a compelling mix of financial incentives (carrots) and legislative requirements (sticks) to provide companies and universities with the right incentives to make the needed investments in cybersecurity.
f. Virtually all countries recognize the importance of education, research, and development for cybersecurity, at all levels—from kids in high school, to seasoned employees working for governments and businesses.
g. Almost no countries look at the Internet of Things in their cybersecurity policies. As mobile phones, cameras, FitBit-like exercise equipment, air conditions, refrigerators, home security systems, medical devices like the Pillcam, automobiles, and more becoming increasingly prevalent, one can expect increased cyber-attacks through vulnerabilities in these devices and the software/apps running on these devices. Today, companies using these devices are focusing on the service they provide (e.g. air-conditioning or monitoring fitness), not the cyber-risks and new vulnerabilities that they are offering to malicious hackers. Without a careful combination of carrots and sticks, it is likely that such companies will continue to unwittingly provide a dramatically larger attack surface to malicious hackers.

1.10 Recommendations

1.10.1 Citizen Education

The first step in shoring up the cyber-defenses of the general public is education. We see that education is a major factor—countries with strong education systems as measured by the UNDP's Human Development Report's various education related measures are generally less vulnerable to cyber-attacks.

Many countries currently offer web sites and courses where their citizens can learn more about cyber-hygiene practices. An excellent example is the Government of Australia that maintains a web site (http://australia.gov.au/ topics/it-and-communications/cybersecurity) which provides simple courses for children, parents, and businesses. Their CyberSmart system educates citizens of all ages. Their ACMAcybersmart channel on YouTube provides a lively set of cyber-videos. Regrettably, only a few of their videos seem to have been heavily viewed (when this chapter was written), while most have just a few hundred views. This suggests that a more robust effort is needed to get the message out. We specifically recommend that:

1. Countries develop educational material relevant to their country's specific needs and host them both on national servers as well as on popular web sites so that others beyond their country can leverage this investment as a shared resource.
   The educational material should include (but certainly shouldn't be limited to):

   a. Video clips of 5 min or less that are posted on YouTube and similar sites;
   b. Audio vignettes played as public service announcements at periodic intervals on appropriate radio stations.

2. All schools should be required to teach cyber-hygiene for at least 1 day per year in much the same way that schools in many countries provide education related to health issues. Countries may designate certain days as "cyber-hygiene" days, days on which all schools are required to teach basic cybersecurity principles to students. Cyber-hygiene testing must be part of the student curricula so that students pay attention to what is being articulated during the cyber-hygiene lectures.
3. Government organizations and companies should include at least 1–2 hours of cyber-hygiene "do's and don'ts" as part of their training programs—both for new employees and as part of re-training and extended learning programs.

1.10.2   Building up a National Cyber-Defense Capability

Different countries have different approaches to dealing with cyber-defense, but the first step is detection. A recent study [2] argues that "a typical zero-day attack lasts 312 days on average and that, after vulnerabilities are disclosed publicly, the volume of attacks exploiting them increases by up to five orders of magnitude." This suggests that countries need to be extremely vigilant in tracking what is going on in networks within their country, and that they maintain constant vigilance on reports of new cyber-vulnerabilities. All of this requires strong cybersecurity training for a core group of cybersecurity professionals (as opposed to the general population). Building a core group of top-notch cybersecurity professionals who work for governments is challenging because top cybersecurity experts tend to work for companies which offer them very high salaries. In addition, as reported by Reuters [11], the US Department of Homeland Security loses even those highly qualified applicants tempted by government jobs to heavy bureaucracy and long wait times. In February 2014, the White House released a set of cybersecurity guidelines that companies can voluntarily follow—but questions remain as to how to track adherence to these guidelines and whether they will be effective [12].

We suggest that countries move forward in accordance with the following recommendations.

1. Provide financial incentives to universities and community colleges to offer cybersecurity courses.

2. Provide financial incentives (e.g. college tuition or credits towards college tuition) to students to adopt careers in cybersecurity in exchange for a certain number of years of government service. This would be much like the ROTC program in the US where students receive free tuition and a small income for 4 years of college in exchange for committing to serve their nation for a few years.
3. Provide competitive salaries for cybersecurity professionals in government and fast-track applicants with demonstrated cybersecurity expertise and training.
4. Provide research grants to universities that are developing algorithms for the real-time identification of cyber-threat as well as for risk mitigation and network hardening.
5. Build national Computer Emergency Response Teams (CERTs) modeled on the US CERT that engage in real-time tracking and identification of cyber-threats.
6. Build partnerships with companies that encourage them to cooperate with the National CERTs through unique financial instruments (e.g. favorable tax treatment) for participating and contributing personnel to national CERTs.
7. Provide financial incentives (e.g. favorable tax treatment or write-offs) for cybersecurity training of company personnel and for sharing educational material related to cybersecurity.

1.10.3   Building Up Timely Cybersecurity Legislation and Regulation

Criminals will always exist, especially when it is possible to take advantage of loopholes both within the laws of individual nations, as well as the loopholes that exist across national boundaries which can be further exacerbated by national and political interests.

Each country not only needs a robust set of laws dealing with cybersecurity, but, perhaps more importantly, a mechanism and agency to enforce, interpret, and adapt those laws to the rapidly evolving IT sector. Specifically:

1. Countries should establish a National Cybersecurity Authority charged with the task of regulating the cyber-sphere, through the interpretation of appropriate laws that are already on the books to new behaviors in cyber-space. Such National Cybersecurity Authorities should also maintain a close watch on nascent technology so that new laws can be passed shortly after new technology is available on the market, rather than several years later.
2. Countries should define a national process to automatically respond to the invention of new cyber-technology that can be used to mount cyber-attacks. Some countries may treat such new technology like weapons technology which needs careful monitoring and regulation, while others may opt for less regulation.
3. Countries should ensure that they have robust laws on the books, together with adequate enforcement authorities, to prosecute individuals for cyber-crime, even if those crimes were committed outside that country's borders.

1.10.4 International Cybersecurity Cooperation

A robust set of international cybersecurity regimes are needed to coordinate cyber-security disputes amongst countries and to regulate cyber-crime. Just as Interpol and Europol help shape multi-national criminal investigations and the European Cyber-Crime Center in The Hague has developed methods to integrate cyber-crime investigations across borders, there is a greater need for a truly international agency charged with reaching multilateral cybersecurity agreements and brokering cyber-security related disputes between nations.

Specifically, we recommend:

- The creation of a G-20 like organization for cybersecurity amongst the most cyber-savvy nations that is robust enough to include major developed economies as well as the major developing economies.
- The creation of an International Cybersecurity Organization, similar to the International Telecommunications Union (ITU), that helps shape multinational agreements around cybersecurity, and provides a mean to enforce them. Specifically, such an organization would place pressure on member nations to adhere to a responsible code of cyber-laws and enforcement within each nation.
- Provide training programs for countries with lower per-capita GDP—as these countries are linked to a higher risk of cyber-vulnerability, there is a greater need to help these countries better develop their cybersecurity capabilities.
- Provide a fund for poorer nations to discharge their responsibilities under such treaty organizations, together with training regimes in which the more cyber-advanced countries share cybersecurity expertise with countries whose host machines may be used to launch attacks on other nations.

1.10.5 Cybersecurity Attack Forecasting

Without a comprehensive understanding of the attacker and the attack surface, countries will not be able to protect their citizens against attack. Countries need to develop a range of new capabilities.

a. *Understanding the Attacker.* Today, most countries are in a reactive mode, responding to attacks. However, attacks by adversaries are driven by a variety of motives. These include situations where one nation-state might attack another in the cyber-domain in order to express dissatisfaction about some policies or statements or actions by the targeted nation. Or the attacks might merely be a criminal attack, aimed at stealing personal information or extorting money from the victim. Today, we have few models of attacker motives. While such models have recently been developed in the domain of counter-terrorism [13, 14], no such predictive models exist for significant cyber-attacks.

b. *Understanding the Victim.* Just as an old lady walking alone during late night is a more likely victim of a mugging than a robust young man doing the same, some host machines are more vulnerable to cyber-attacks than others. There has been relative little work in this area though recent work has just started [15]— Chap. 4 provides some initial work on understanding the victim.

c. *Predicting the Spread and Impact of an Attack.* Classical epidemiology tells us that some diseases spread much more rapidly than others. Enterprises and countries cannot assign resources to protect against a particular piece of malware without a detailed understanding of how many hosts a particular piece of malware will affect, and the impact of such infections on the functioning of the enterprise. Today, there are almost no predictive models of spread (i.e. number of hosts likely to be infected by a piece of malware), though [16] does so by building on top of epidemiological models [17]. Much more work along these lines is needed in order to suggest cybersecurity priorities and investments to national cybersecurity officials.

d. *Understanding the Evolution of the Attack Surface.* The bulk of the work in cybersecurity today has focused on understanding the attack surface in order to identify vectors that the attacker might use. However, to date, countries have not invested significant amount of money in understanding how the attack surface will evolve over time as new technological innovations (e.g. new communication protocols, new financial models, new "internet of things" based products) are introduced into the market. Countries must invest heavily in understanding the new cyber-vulnerabilities introduced by such products, business models, and services, and forecast how these vulnerabilities might be exploited by malicious hackers in the future.

1.11 Conclusion

This book contains the results of a detailed 2-year data set about hosts around the world who were protected by Symantec's anti-virus products such as Norton Anti-Virus. All hosts participating in the study did so on a voluntary, opt-in basis, and all personally identifying information was stripped from the data by Symantec before the data was made available to us.

We looked at two major measures of a country's cyber-vulnerability:

- The average number of attacks per machine in that country;
- The fraction of machines in that country that were attacked.

Our results were limited to the 44 countries for which we had data on at least 500 machines in each of the 2 years. Of these 44 countries, the most cyber-vulnerable nations in the world are India, S. Korea, Saudi Arabia, China, Malaysia, and Russia.

At the same time, the least cyber-vulnerable countries in the world are the Scandinavian countries (Norway, Denmark, Finland, and Sweden), Switzerland, and Germany.

Countries that are frequently thought of as hugely cyber-vulnerable such as the US and UK are in fact amongst the safer countries in the world from a cybersecurity perspective, coming in ranked around 10–12 in terms of cyber-invulnerability.

Last but not least, our results also shed light on relationships between software piracy, per-capita GDP, human development index, health index, educational systems, and the cyber-vulnerability of countries. Unsurprisingly, countries with low per-capita GDP, low HDI, low health indexes, and weaker education systems have higher cyber-vulnerability. Moreover, countries with high software piracy rates also have high cyber-vulnerability—however, our statistical findings show that though this relationship is statistically sound at a univariate level, a multivariate analysis reveals that other variables better explain cyber-attacks on host machines.

Nonetheless, we caution the reader that correlation is not causation. We have no way (nor do much more powerful agencies like the World Bank, UNDP, ITU and others) of running a controlled experiment across countries (or even just one country) in which we vary the values of variables like per-capita GDP, HDI, and assess the impact of these variations on the cyber-vulnerability of these countries.

We treat software piracy slightly differently than these variables because companies can unilaterally (or almost unilaterally) lower prices in extremely cyber-vulnerable countries with low GDPs. This would likely increase the purchase of legitimate software without embedded exploits, which would increase market share in those countries (potentially offsetting some of their losses), while decreasing cyber-vulnerability for all. We urge major software companies to try such an experiment in controlled and perhaps small-scale settings to see where this works and where it does not, and we also urge software industry groups to consider doing this in the most cyber-vulnerable countries. As such countries also largely happen to be poor, this would be a valuable public good as well.

References

1. Mezzour G, Carley KM, Carley LR (2015) "An empirical study of global malware encounters." In: *Proceedings of the 2015 Symposium and Bootcamp on the Science of Security*. ACM, 2015
2. Bilge L, Dumitras T (2012) Before we knew it: an empirical study of zero-day attacks in the real world. In: ACM Conference on Computer and Communications Security, pp 833–844
3. Symantec Corp (2013) Four years of DarkSeoul cyberattacks against South Korea continue on anniversary of Korean War. Technical report http://www.symantec.com/connect/blogs/four-years-darkseoul-cyberattacks-against-south-korea-continue-anniversary-korean-war
4. Gantz JF et al (2013) International Data Corporation. The dangerous world of counterfeit and pirated software. Technical report http://www.microsoft.com/en-us/news/download/presskits/antipiracy/docs/idc030513.pdf
5. Federal Bureau of Investigation (2013) Consumer alert: Pirated software may contain malware. Technical report https://www.fbi.gov/news/stories/2013/august/pirated-software-may-contain-malware

6. Business Software Alliance. Shadow market—2011 BSA global software piracy study, 2012. 9th edition, http://globalstudy.bsa.org/2011/downloads/study_pdf/2011_BSA_Piracy_Study-Standard.pdf
7. McAfee Labs (2013) McAfee threats report: First quarter 2013. Technical report http://www.mcafee.com/us/resources/reports/rp-quarterly-threat-q1-2013.pdf
8. McAfee Labs (2014) McAfee threats report August 2014. Technical report. http://www.mcafee.com/us/resources/reports/rp-quarterly-threat-q2-2014.pdf
9. Symantec Corp (2014) Symantec internet security threat report 2014 appendix. Technical report http://www.symantec.com/content/en/us/enterprise/other_resources/b-istr_appendices_v19_221284438.en-us.pdf
10. Symantec Corp (2014) Symantec internet security threat report 2014. Technical report http://www.symantec.com/content/en/us/enterprise/other_resources/b-istr_main_report_v19_21291018.en-us.pdf.
11. Chiacu D (2014) Homeland security struggles to tempt, retain cyber talent. In: Reuters, Apr 26, 2014 http://www.reuters.com/article/2014/04/26/us-usa-cybersecurity-dhs-idUSBREA3P05O20140426
12. Hennessey K, O'Brien C (2014) Cybersecurity guidelines for companies are unveiled by white house. In: Los Angeles Times, Feb. 12, 2014 http://articles.latimes.com/2014/feb/12/business/la-fi-cyber-security-20140213
13. Subrahmanian VS, Mannes A, Sliva A, Shakarian J, Dickerson J (2012) Computational Analysis of Terrorist Groups: Lashkar-e-Taiba, Aug 2012, Springer, Switzerland
14. Subrahmanian VS, Mannes A, Roul A, Raghavan RK (2013) Indian Mujahideen: Computational Analysis and Public Policy, Springer, Nov 2013
15. Ovelgonne M, Dumitras T, Prakash BA, Subrahmanian VS, Wang B (2015) Understanding the Relationship between Human Behavior and Susceptibility to Cyber-Attacks: A Data-Drive Approach, submitted to a technical journal, Feb 2015. Currently being revised
16. Kang C, Park N, Prakash A, Serra E, Subrahmanian VS (2015) Ensemble Models for Data-Drive Prediction of Malware Infections, to appear in: Proc. 9th ACM International Conference on Web Science and Data Mining (WSDM), San Francisco, Feb 2016
17. Matsubara Y, Sakurai Y, van Panhuis WG, Faloutsos C (2014) FUNNEL: automatic mining of spatially coevolving epidemics. In: *Proceedings of the 20th ACM SIGKDD international conference on Knowledge discovery and data mining* pp 105–114 ACM, Aug 2014

Chapter 2
Types of Malware and Malware Distribution Strategies

Using data from the Identity Theft Supplement to the National Crime Victimization Survey [1], the US Department of Justice estimates that approximately 7% of all Americans over the age of 16 have been victims of identity theft. Over 45% of the victims of identity theft reported over 6 months of stress resulting from the incident. The situation in other developed economies is similar—according to CIFAS, the UK's Fraud Prevention Service, 108,500 people had their identity stolen in the UK in 2013 [2].

The stress placed by these statistics—both on the victims of identity theft and those who fear it—is substantial. It is not helped by reports in late 2014 of highly targeted attacks on consumers such as the *Darkhotel* "espionage campaign" reported by Kaspersky Labs [3] in which a sophisticated ring of cyber-criminals target individuals who are wealthy enough (and presumably influential enough) to stay at high-end luxury hotels. When these individuals access the hotel's Wi-Fi network, they are asked to update a piece of software which induces most such individuals to download a malicious piece of code onto their devices. Once on the unsuspecting victim's device, *Darkhotel* runs in the background, downloading, installing, and deleting at will, advanced software such as keystroke loggers, Trojans, and various malware designed for data and information theft.

Likewise, Kaspersky Labs [4] reports that a 2013–2014 joint study with Interpol found that approximately 20% of Android devices protected by Kaspersky Labs detected attacks on those devices, suggesting that perhaps 20% of all Android devices are thus targeted. This is a very steep rise on numbers reported from just 1 year earlier.

All of this makes consumers worldwide fearful of malware and its ability to derail their finances, and eventually their lives. Businesses are equally nervous about the asymmetric nature of the threat posed by malware developers and nation states. Businesses are worried both about *insider threat* (where insiders steal corporate secrets) [5] as well as commercially motivated external threats [6]. For example, the FBI issued a warning in November 2013 describing a piece of malware that over-writes hard drives, thus destroying corporate data. According to an earlier

© Springer International Publishing Switzerland 2015   33
V.S. Subrahmanian et al., *The Global Cyber-Vulnerability Report*, Terrorism,
Security, and Computation, DOI 10.1007/978-3-319-25760-0_2

report by the security firm Mandiant, "APT1 has systematically stolen hundreds of terabytes of data from at least 141 organizations, and has demonstrated the capability and intent to steal from dozens of organizations simultaneously" [7, p. 5]. They identify APT1 as being located in the same location as the People Liberation Army's Unit 61398 in the Pudong area of Shanghai.

Governments are constantly fearful of cyber-espionage attacks by foreign states. For years, PLA's Unit 61398 was viewed in the US and EU as being the poster child for unwarranted and unethical cyber-espionage [7–9] (though this has been dialed down since allegations of widespread cyber-espionage by the US government hit the news in the summer of 2013 following the revelations of Edward Snowden). Nonetheless, the FBI recently warned [10] that an even more deadly adversary codenamed *Axiom* within the PLA has been stealing intellectual property from US companies, engaging in cyber-espionage, and in targeting Chinese dissidents. The Chinese government has steadfastly denied all such allegations. But we should note that there are allegations that the US too maintains a stock of zero-day attacks and does not disclose all vulnerabilities it has discovered in software to the software vendors involved [11].

In short, there are huge numbers of zero day attacks and advanced persistent threats being developed by a number of actors ranging from individual hackers to criminal groups to nation states. In the rest of this chapter, we provide the briefest insights into the different types of malware that are exploited by many of these entities, as well as some of the mechanisms used to distribute these attacks.

2.1 Types of Malware

We describe six type of malware: Trojans, viruses, worms, spyware, adware, and misleading software. These are the types of malware that we studied in the WINE dataset from Symantec. Of course, there are many other kinds of malware as well, and we will summarize some of these other types of malware toward the end of this section.

2.1.1 Trojans

A Trojan is a hidden threat, much like the famed Trojan horse left by Odysseus on the shores of Troy.

Simply put, a Trojan consists of two parts—a server side that runs on an attacked host and a client piece that runs on the attacker's console. The server code (usually kept very small in size, no more than a few KBs) is dispatched to the victim via some malware distribution method. We will describe several malware distribution methods in Sect. 2.2 including phishing attacks, drive-by-downloads, and so forth. In a simple setting, the attacker sends the victim a file that contains the server code (e.g. an image or a PDF large enough in size that the server size is miniscule when compared to the overall file size). When the user double clicks the attacked file, it launches the "server" program embedded in the infected file. The server usually runs in stealth mode and is not easily visible to the user and/or to the file manager. At this stage, the server code in the infected file can establish contact with the attacker's client code in one of many ways. One simple way is through a reverse connection in which the server code has the IP address from which the attacker wants to control the victim's computer. But much more sophisticated reverse connection methods also exist. Once launched, the server program contacts the client side code from whose console, the attacker can now take control of the victim's program. He can install new programs on to the victim's computer (e.g. keyloggers), he can read every single file on the victim's computer (e.g. credit card and banking information, personal identity information), and more. In effect, he can control the victim's computer using his keyboard from a remote location.[1]

In some cases, Trojans are very explicit and make few attempts to stay "below the radar". They take overt control of the victim's machine. The more dangerous situation, however, is when the Trojan stays below the radar and operates for extended periods of time in stealth mode with the victim unaware that his data (or his company's data) is being siphoned off by an unscrupulous attacker.

One example of a dangerous Trojan is the Zeus3 malware which was downloaded onto victims' computers through infected advertisements that may be present on various web sites. When these ads are viewed by the victim, the Trojan is downloaded onto the victim's computer. The Trojan then waited till the user visited his online bank. By observing his credentials when he logs in, it is able to siphon off a large sum of money from victims' bank accounts.

Another interesting Trojan, Obad.a infects Android devices [12, 13] by first sending potential victims an infected link (or a spam message). When the victim clicks the link, he downloads the Trojan server onto his device which immediately reaches out to his entire contact list, urging them to click on the link as well. The Trojan spreads in this way, infecting a large number of people. Unlike most Trojans, this one uses a botnet to control the spread of the Trojan.

Another Trojan, CryptoLocker, encrypts user files on a machine and demands a ransom in exchange for decrypting the file.

Most intriguing is the recent report of the *Regin* virus in a report released by Symantec [14] and Kaspersky Lab [15]. Regin is primarily used for espionage and intelligence gathering. According to Symantec [14, p. 6], 48% of infection attempts target private individuals, 28% of infection attempts target telecommunication companies, and the rest is split between the hospitality industry (hotels), airlines, energy sector, and researchers. The main affected countries are Russia, Saudi Arabia, Mexico, Ireland, Afghanistan, India, Iran, Belgium, Austria, and Pakistan—with Russia and Saudi Arabia the biggest targets. Interestingly, [15] reports that Regin also compromises GSM networks and collects data about the physical networks

---

[1]The ability to control an infected host from a remote machine is a featured shared by different types of malware, not just Trojans.

used by telecoms. It also collects administrative login data that allows it to manipulate the networks. According to the German newspaper *Der Spiegel* [15], Regin is a joint effort of the US National Security Agency and the UK's GCHQ.

Regin starts with a "dropper" in which the malware is dropped onto a site. Some evidence suggests that Instant Messaging services are used to inject Regin into certain hosts. From there, several complex intermediate steps (including ones involving encryption) are performed before the ultimate payload is revealed. The goal is to steal information from the compromised hosts. In order to evade detection, Regin compromises entities in a country by linking them into an intra-country peer to peer network and then using just one exit point from the country to exfiltrate the data to its creators' location. [15] shows a graphic of how India's President's office and many government institutions were linked into a P2P network with the single entry/exit point from India being a compromised node at an educational institution.

We see therefore that Trojans can vary widely in sophistication, ranging from software that is likely designed by teams of dedicated hackers working for a nation state, to individual hackers or hacker collectives taking known code and modifying it. Because this book focuses primarily on infection attempts on consumer hosts as opposed to government or business hosts, we believe that most of the Trojans described in this book are in the second category.

2.1.2   Worms

A worm is a piece of malware that can independently spread through a network by exploiting vulnerabilities in existing software to compromise a system. Worms may spread through networks in a variety of ways. For instance, worms may spread through a network by using email to infect other computers, or by using other file transfer protocols to copy themselves onto other computers.

Worms may carry a payload. While some worms may do nothing other than spread from one computer to another (just using up bandwidth and slowing down a network), others may do dangerous things like delete files on a machine and encrypt files (so that the owner of the file has to pay a ransom in order to be able to decrypt his files).

[16] provides a detailed taxonomy of worms based on six factors.

- *Targeting*. This refers to the mechanism used by the worm to target potential victims. Commonly used targeting mechanisms include scanning the network for vulnerable hosts, using specified lists of targets, using a "metaserver" (which is a list of periodically updated vulnerable servers) that the worm periodically queries to find new targets, and topological worms that discover the structure of a network in order to identify new targets, and "passive" worms that lie in wait for a target.
- *Distribution Mechanisms*. Worms might spread in three ways. Self-carried worms spread independently (e.g. topological worms and worms that spread by

scanning through a network). Second-channel worms spread via an auxiliary communication channel such as remote procedure calls. Embedded worms spread by embedding themselves within a standard channel of communication.

- *Activation Mechanism.* Worms may be activated either by an explicit human action (e.g. via an infected email), an explicit human activity that is recognized by the worm, triggering it, or by a injecting themselves into part of a scheduled process on a host.

In general, topological worms and worms that spread autonomously by scanning can be incredibly fast. Notorious computer worms include:

- *Stuxnet* [17–19] is perhaps the best known example of a worm in recent years. Detected in 2013 by security vendor Kaspersky Labs [17], and reportedly launched by Israeli and US intelligence [20], Stuxnet was signed with certificates stolen from two Taiwanese software manufacturers, making it appear to be authentic and reliable. Stuxnet was targeted at Iran's Natanz nuclear enrichment facility. Though Stuxnet code infected computers in many nations, it is reported [18] that it did not adversely impact any SCADA systems other than those at Natanz. Stuxnet worked via an initial socially engineered attack in which a memory stick infected with Stuxnet was introduced. The worm spread rapidly. When infecting a host, Stuxnet first checked to see if it was a particular kind of Siemens device often used in nuclear facilities. If it was, a dropper program dropped malicious code into the main() program loop of the Siemens controller. The malicious code included several variants targeted at the specific type of controller.
- *Mydoom* [21] appears with a message in emails, prompting (mostly Windows) users to click upon an attachment, upon which their machine is infected. Different versions of Mydoom carry different payloads, one of which is the installation of a backdoor on the victim machine that allows the machine to be remotely controlled. Mydoom is believed to have used up a huge amount of Internet bandwidth when it first hit the internet in 2004.
- *Conficker* [22, 23] exploits a vulnerability in the Windows operating system to infect a host—and does this by a combination of random scans of nodes as well as neighborhood scans (i.e. scanning neighbors of infected nodes), though the latter is reported to be the dominant mode of infection [22]. Conficker was sophisticated enough to update itself dynamically and also evade signature-based anti-virus detection tools.

2.1.3 Viruses

Unlike worms, that spread independently, viruses spread by attaching themselves to another program or to files (e.g. PDF or image files). For example, a virus embedded in a PDF or JPEG file may spread when that file is opened. Some viruses also exist in the boot sector of a computer hard drive, thus executing automatically when a boot operation takes place.

Because legitimate programs and files have well known sizes, viruses that attach themselves to such "entities" may take steps to hide any increase in size. One way to hide is by copying themselves into unused space in a file or program. Another way to hide is by intercepting requests to obtain data about the program or file and returning results that appear normal and obfuscate the presence of the virus. In order to hide from "signature based" scanners used by many anti-virus companies (a signature is just a fragment of code), viruses can mutate, making their code look different. Rates of mutation vary from one virus to another.

It is unfortunate that in common parlance, the word "virus" has been collectively used to describe all kinds of malware including worms, Trojans, and viruses as described above.

2.1.4 Other Forms of Malware

Other forms of malware include "misleading software" and "spyware".

We use the term "misleading software" to describe software that pretends to be something legitimate, when in fact it is really a piece of malware. Examples of misleading software include fake anti-virus programs, fake media players, and fake hard disk recovery programs.

Fake anti-virus software use social engineering to make users believe their system is infected with a virus. A free Anti-Virus software is offered and shows fake infection results when it is downloaded and run. Then, the user receives an offer to upgrade the software (for a fee) to remove the supposedly existing infection. Another type of misleading software, sometimes pretends to present messages from a law enforcement agency. The user is accused of a crime and the payment of a fine is requested. Rajab et al. [24] and Stone-Gross et al. [25] provide further details on misleading software.

A related type of malware is ransomware which encrypts files on the host of a victim and demands a ransom [16]. As many users never create regular backups, a victim can only regain access to his/her files after paying the requested ransom to the attacker.

Spyware is code that enables a third party to spy on a host. Spyware has been used for a variety of purposes including identity theft and theft of personal data, spying on online activities of individuals (e.g. spouses) and watching users' online activities.

2.2 Malware Distribution

We now come to the important topic of malware distribution. Though malware distribution can occur in many different ways, we focus on four of them: drive-by-downloads, email, network intrusion, and social engineering.

2.2.1   Drive-by-Downloads

The main characteristic of drive-by attacks is that the user unknowingly downloads a malicious file while browsing the web. Some component of the web browser or one of its plug-ins (e.g. those to display PDF or Flash files), processes the malicious file. Malicious Web code (e.g. JavaScript) exploits vulnerabilities in browsers and causes a file to be downloaded and executed. Because of the availability of various injection techniques, as described below, the malicious code may be present on Web pages that are popular and otherwise benign.

Drive-by attacks require a victim to visit a website that contains attack code. Building on the injection strategies in Provos et al. [26], we can categorize injection strategies into four categories:

- They can post malicious code as a part of a submission to a user contributed website that does not carefully ensure that user inputs are malware-free.
- They can include malicious code into ads and pay unsuspecting and/or careless ad networks to deliver the ad to their client websites.
- They can provide malicious widgets like stats counter. Websites that include the widgets deliver the code to their visitors.
- Adversaries can try to get control of the web server of a benign website and add their code to it.

Though shady web sites (e.g. porn sites) seem to pose a greater risk of drive-by attacks, visiting only large and/or popular web sites does not entirely mitigate the risk of being victimized. For example, it is reported that around the turn of the year 2013–2014, visitors to Yahoo sites were served ads from Yahoo's ad network [27] that were infected with malware.

2.2.2   Email

As in the case of drive-by attacks, e-mail attacks can exploit vulnerabilities in the e-mail software or in the libraries that the e-mail software uses (e.g. to display images or to display Word or PDF files). When the email software downloads a message and displays it, a manipulated embedded media object exploits a vulnerability and causes the execution of the malicious code.

2.2.3   Network Intrusion

While drive-by attacks and email attacks require that the victim initiate communication with a remote host, network intrusion attacks are initiated by the attacker. Victim hosts run programs that process incoming data on several layers of the protocol stack. Manipulated data packages can exploit vulnerabilities and take over control of a host.

2.2.4 Social Engineering

Socially engineered attacks exploit weaknesses of humans rather than weaknesses of software. Users are manipulated into running malicious binaries.

For example, users are made to believe there is malware on their computer and offered a free Anti-Virus software (compare Sect. 2.1.4). As this malware distribution strategy does not exploit any technical vulnerability, the hurdle to overcome is that of public awareness.

A well-known example of social engineering is the Koobface attack (https://nakedsecurity.sophos.com/koobface/), which would identify the Facebook accounts accessed from the infected computers and post messages using those accounts. This leveraged the established trust between those users and their Facebook friends.

In addition to attacks that rely entirely on it, social engineering is also involved (to varying degrees) in most other distribution strategies as well. For example, malware distribution involving email may exploit vulnerabilities of software other than email software by using social engineering to make the user open an attached files. For example, an adversary may send fake invoices. When a user opens the unexpected invoice to see what it is about, malicious code gets executed.

2.2.5 Downloaders

All the methods to distribute malware discussed above are initial attacks. A common way to distribute malware is to install a malware downloader using one of the initial attack methods discussed above. Once installed, a downloader downloads and installs additional malware on a previously compromised host.

A downloader system can be regarded as a special type of botnet where the downloader is a bot that specializes in retrieving and installing malware. Technical details about how downloader networks operate can be found in [28] and [29].

2.3 Business Models

In this section, we present an overview on the most common business models of the underground economy.

Making money in the underground economy is a multi-step process. The process starts with the identification of vulnerabilities of operating systems and pieces of software (Exploit-as-a-service (EAAS) [30]) and ends with a transfer of funds, e.g. through dubious payment processors for credit cards [25].

As in the case of the traditional economy, the cyber-crime economy adopted a division-of labor model where individuals or organizations specialize in one part in the value chain.

A comprehensive service in the dark economy is Pay-Per-Install (PPI) [28]. A PPI provider takes over the complex task of identifying and exploiting vulnerabilities (or buys these from other service providers) and installs downloaders on compromised hosts.

Other services include solving CAPTCHAs (Completely Automated Public Turing Test to tell Computers and Humans Apart) for a variety of purposes of dubious legality. Solving such CAPTCHAs may support posting advertisements for malicious web sites and online message boards, creating accounts at free e-mail services [31], and repacking malware to prevent signature-based identification through anti-virus software services to promote malicious websites [25].

Some money-making methods used in this underground economy are listed below.

2.3.1 Click Fraud

Cost-per-click (CPC) is a common compensation method in online advertising. The website displaying an ad gets paid not for displaying the ad, but for every click that takes a visitor to the advertiser's website. Click fraud can work in two ways. First, there are owners of websites who wants to use so called click-bots to increase the clicks on ads on their website to increase their own ad revenue. In the same vein, there are organizations who want to increase the spending of their competitors and use click-bots to click on their competitor's ads.

Another type of click fraud is fraudulent search engine optimization (blackhat SEO) [32, 33]. Search engines rank their list of result based, in part, on result entries that users clicked on in the past. Here, click fraud malware is used to fool search engines into believing a website is more popular than it actually is.

2.3.2 Keyloggers

Keyloggers collect personal information like bank account or credit card data and email credentials. This type of information is a marketable product in the underground economy [34, 35] and can be used in different types of fraud schemes or to send spam.

2.3.3 Spam

Unsolicited bulk email is the most common form of spam. But similar messages are also sent to message boards on the web, to social media (e.g. YouTube) and to social network (e.g. Facebook) sites. Spam is often used to deceive victims into buying worthless or dangerous products (e.g. counterfeit prescription drugs [36]).

However, spam is also used to distribute malware (compare Sect. 2.2) and keep the cycle of infections going. Some spam-based malware networks exfiltrate address books from compromised hosts to build email databases [37].

2.4   Cross-Country Studies

The security literature includes thorough studies analyzing malware offenders through honeypots [37], scanning network traffic [38], or by milking PPI distribution servers [28]. Studies about the victims of malware are, however, rather rare. We summarize below the cross-country results those victim-centric studies revealed.

Caballero et al. [28] infiltrated PPI networks and studied their behavior. They "milked" PPI services by using software that resembled the original downloader of a PPI service to retrieve the binaries it distributes. By accessing the PPI service with IPs from different geographical origins they were able to study the behavior of these services across different countries. They observed that while most malware was not uniformly distributed across countries, most malware families did have geographic preferences and specifically targeted either the United States or Europe. They attribute the country preferences to (1) the varying pay-per-install costs they found at dark marketplaces for such services ($100–180 for 1K US or UK host, $20–160 for 1K hosts in other European countries), and (2) the need to customize some attacks. Success in stealing credit card data or advertising and selling fake anti-virus software depends on the geography of the victims. Credit cards are not widely used in all parts of the world and payment methods need to fit local systems. Selling fake anti-virus software most likely works best when the advertising message is in the native language of the victim.

Other business models do not require a country-specific approach. In order to send spam, any host connected to the internet is good — with the exception of countries that might be blacklisted or trigger spam filters.

Carlinet et al. [38] conducted a detailed study of how behaviors affect vulnerability to malware by malware. They analyzed the network traffic of several thousand ADSL-customers of the French network operator Orange to identify risky types of applications. The presence of malware was inferred by running a signature-based intrusion detection system (IDS) on the traffic data. They found that web and streaming usage is a risk factor while this is not the case for other types of applications like peer-to-peer and chat applications. This result is not surprising, given that browser-based drive-by attacks is the most popular malware distribution approach.

L'evesque et al. [39] ran a field experiment with 50 persons. They installed software on laptops to monitor web browsing behavior and malware infection and then handed them out to their subjects. Data collected over a period of 4 months indicates that higher computer literacy is positively correlated to malware infections. This is a counterintuitive result, that we also saw in our much larger study of millions of hosts in Symantec's WINE database (see Chap. 3). L'evesque et al. [39] does not analyze how much the factors they analyze influences the infection risk.

Shin et al. analyzed the victims of botnets [40]. They collected IP addresses of hosts infected by three different botnets and analyzed the number of infected networks (/24 IP address space). The network level aggregation has been conducted to account for dynamic IP assignment. One of the analyzed botnets uses a self-propagating approach (type I) and two use a distributed malware-propagation approach (type II). Shin et al. observed that most countries have similar share of type I and type II attacks. But some countries like China have a much higher share of type I infected networks than type II infected networks. They assume network management policies could be a reason for this. The percentage of infected subnets of a country that [41] computed gives a totally different results than the percentage of infected hosts we observed (see Chap. 3). We believe their analysis suffers from the following flaws:

- Because of dynamic IP addresses, their data does not reveal how many hosts belong to a/24 subnet.
- Differences in the number of server farms or private hosts with static IP address influence the average number of hosts per subnet.
- Different ratios of desktop hosts/servers bias the results as well.

2.5 Conclusion

In short, we see that there are currently several types of malware available in the wild and are distributed to potential victims through a number of sophisticated methods. Moreover, they support business models that range from espionage and theft by nation states, to common criminals who are motivated by economic greed.

Though we have only described a few specific types of malware such as Trojans, worms, viruses, misleading applications, and ransomware to name a few, many pieces of malware can often be pieced together into complex "botnets" of malicious programs working together across networks in order to achieve their ends. Complex malware such as Stuxnet and Regin are believed to have been designed and executed by nation states.

Malware is distributed by a variety of methods ranging from spam and malicious web sites on the one hand, to infected attachments that are mailed to potential victims.

Malware supports a variety of business models. Excluding espionage and data theft at the nation state level, malware is used to promote web sites, generate spam, generate fraudulent clicks to increase revenues of web sites (or increase cost of rival web sites), and promote sales of fake products such as fake anti-virus and fake disk cleanup packages.

References

1. Harrell E, Langton L (2014) Victims of Identity Theft 2012, US Bureau of Justice Statistics, http://www.bjs.gov/content/pub/pdf/vit12.pdf, retrieved Dec 3 2014
2. CIFAS (2014) Is Identity Fraud Serious, https://www.cifas.org.uk/is_identity_fraud_serious, retrieved Dec 3 2014
3. Kaspersky Labs Virus News (2013) Kaspersky Lab sheds light on "Darkhotels", where business executives fall prey to an elite spying crew, Nov 14 2013, http://www.kaspersky.com/about/news/virus/2014/Kaspersky-Lab-sheds-light-on-Darkhotels-where-business-executives-fall-prey-to-an-elite-spying-crew, retrieved Dec 3 2014
4. Kaspersky Labs (2014) Kaspersky Lab & INTERPOL Report: Every Fifth Android User Faces Cyber-Attacks, Oct 6 2014, http://www.kaspersky.com/about/news/virus/2014/Every-Fifth-Android-User-Faces-Cyber-Attacks, retrieved Dec 3 2014
5. Azaria A, Richardson A, Kraus S, Subrahmanian VS (2014) Behavioral Analysis of Insider Threat: A Survey and Bootstrapped Prediction in Imbalanced Data, accepted for publication in IEEE Transactions on Computational Social Systems, vol 1(2) pp 135-155
6. Halleck T (2014) FBI Says Cyber Attacks On US Businesses Have Followed Sony Hack, International Business Times, Dec 1 2014, http://www.ibtimes.com/fbi-says-cyber-attacks-us--businesses-have-followed-sony-hack-1731670, retrieved Dec 3 2014
7. Mandiant Corporation (2013) APT1Exposing One of China's Cyber Espionage Units, http://intelreport.mandiant.com/Mandiant_APT1_Report.pdf, retrieved Dec 3 2014
8. Brenner J (2011) America the Vulnerable: Inside the New Threat Matrix of Digital Espionage, Crime, and Warfare. Penguin
9. Clarke RA, Knake RK (2011) Cyber war. HarperCollins
10. Nakashima E. (2014) Researchers identify sophisticated Chinese cyberespionage group, Oct 28 2014, http://www.washingtonpost.com/world/national-security/researchers-identify-sophisticated-chinese-cyberespionage-group/2014/10/27/de30bc9a-5e00-11e4-8b9e-2ccdac31a031_story.html, retrieved Dec 3 2014
11. Zetter K (2014) U.S. Gov Insists It Doesn't Stockpile Zero-Day Exploits to Hack Enemies, Nov 17 2014, Wired, http://www.wired.com/2014/11/michael-daniel-no-zero-day-stockpile/, retrieved Dec 3 2014
12. Kaspersky Labs (2013) First ever case of mobile Trojan spreading via 'alien' botnets, Sep 5 2013, http://www.kaspersky.com/about/news/virus/2013/first_ever_case_of_mobile_Trojan_spreading_via_alien_botnets, retrieved Dec 3 2014
13. Unuchek R (2013) The Most Sophisticated Android Trojan, June 6 2013, http://securelist.com/blog/research/35929/the-most-sophisticated-android-trojan/, Retrieved Dec 03 2013
14. Symantec (2014) Regin: Top-tier espionage tool enables stealthy surveillance, Nov 24, 2014 http://www.symantec.com/content/en/us/enterprise/media/security_response/whitepapers/regin-analysis.pdf, retrieved Dec 3 2014
15. Kaspersky Lab (2014) Regin: a malicious platform capable of spying on GSM networks, Nov 24 2014, http://www.kaspersky.com/about/news/virus/2014/Regin-a-malicious-platform-capable-of-spying-on-GSM-networks, retrieved Dec 03 2014
16. Weaver N, Paxson V, Staniford S, Cunningham R (2003) A taxonomy of computer worms. In: Proceedings of the 2003 ACM Workshop on Rapid Malcode, WORM'03, pp 11–18, NY, USA
17. Kushner D (2013) The real story of Stuxnet. IEEE Spectrum, 50(3), 48–53
18. Langner R (2011) "Stuxnet: Dissecting a cyberwarfare weapon." IEEE Security & Privacy, vol. 9(3)49–51
19. Matrosov A, Rodionov E, Harley D, Malcho J (2010) Stuxnet under the microscope. ESET LLC report
20. Nakashima E, Warrick J (2012) Stuxnet was work of US and Israeli Experts, Officials Say, June 12 2012, Washington Post http://www.washingtonpost.com/world/national-security/stuxnet-was-work-of-us-and-israeli-experts-officials-say/2012/06/01/gJQAlnEy6U_story.html, Retrieved Dec 16 2014

21. Sung AH, Xu J, Chavez P, Mukkamala S (2004) Static analyzer of vicious executables (save). In: IEEE Computer Security Applications Conference, Dec 2004. 20th Annual, pp 326–334

22. Shin S, Gu S, Gu G (2010) Conficker and beyond: a large-scale empirical study. In: ACM Proceedings of the 26th Annual Computer Security Applications Conference, pp 151–160

23. Porras P (2009) Inside risks reflections on Conficker. In: Communications of the ACM, 52(10)23–24

24. Abu Rajab M, Ballard L, Mavrommatis P, Provos N, Zhao X (2010) The nocebo effect on the web: An analysis of fake anti-virus distribution. In: Proceedings of the 3rd USENIX Conference on Large-scale Exploits and Emergent Threats: Botnets, Spyware, Worms, and More, LEET'10, Berkeley, CA, USA, USENIX Assoc

25. Stone-Gross B, Abman R, Kemmerer RA, Kruegel C, Steigerwald DG, Vigna G. The underground economy of fake antivirus software. In: Schneier B (ed) Economics of Information Security and Privacy III, Springer, New York, pp 55–79

26. Provos N, McNamee D, Mavrommatis P, Wang K, Modadugu N (2007) The ghost in the browser: Analysis of web-based malware. In: Proceedings of the 1stWorkshop on Hot Topics in Understanding Botnets (HotBots)

27. Fox IT (2014) http://blog.fox-it.com/2014/01/03/malicious-advertisements-served-via-yahoo/.

28. Caballero J, Grier C, Kreibich C, Paxson V (2011) Measuring pay-per-install: The commoditization of malware distribution. In: Proceedings of the 20th USENIX Security Symposium, San Francisco, CA, USA

29. Rossow C, Dietrich C, Bos H (2013) Large-scale analysis of malware downloaders. In Flegel U, Markatos E, Robertson W (eds) Detection of Intrusions and Malware, and Vulnerability Assessment, vol 7591 of Lecture Notes in Computer Science. Springer, Berlin Heidelberg, pp 42–61

30. Grier C, Ballard L, Caballero J, Chachra N, Dietrich CJ, Levchenko K, Mavrommatis P, McCoy D, Nappa A, Pitsillidis A, Provos N, MZ Rafique, Abu Rajab M, Rossow C, Thomas K, Paxson V, Savage S, Voelker GM (2012) Manufacturing compromise: The emergence of exploit-as-a-service. In: Proceedings of the 2012 ACM Conference on Computer and Communications Security, CCS '12, pp 821–832, New York, NY, USA

31. Namestnikov Y (2009) The economics of botnets. Technical report, Kaspersky Labs, https://www.securelist.com/en/downloads/pdf/ynam_botnets_0907_en.pdf

32. John JP, Yu F, Xie Y, Krishnamurthy A, Abadi M (2011) deseo: Combating search-result poisoning. In: Proceedings of the 20th USENIX Conference on Security, SEC'11, pp 20–20, Berkeley, CA, USA, USENIX Assoc

33. Lu L, Perdisci R, Lee W (2011) Surf: Detecting and measuring search poisoning. In: Proceedings of the 18th ACM Conference on Computer and Communications Security, CCS'11, pp 467–476, New York, NY, USA

34. Franklin J, Paxson V, Perrig A, Savage S (2007) An inquiry into the nature and causes of the wealth of internet miscreants. In: Proceedings of the 14th ACM Conference on Computer and Communications Security, CCS '07, pp 375–388

35. Holz T, Engelberth M, Freiling F (2009) Learning more about the underground economy: A case-study of keyloggers and dropzones. In: Backes M and Ning P (eds) Computer Security — ESORICS 2009, vol 5789 of Lecture Notes in Computer Science, Springer Berlin Heidelberg, pp 1–18

36. McCoy D, Pitsillidis A, Jordan G, Weaver N, Kreibich C, Krebs B, Voelker GM, Savage S, Levchenko K (2012) Pharmaleaks: Understanding the business of online pharmaceutical affiliate programs. In: Proceedings of the 21st USENIX Conference on Security Symposium, Security'12, pp 1–1, Berkeley, CA, USA, USENIX Assoc

37. Polychronakis M, Mavrommatis P, Provos N (2008) Ghost turns zombie: Exploring the life cycle of web-based malware. In: Proceedings of the 1st Usenix Workshop on Large-Scale Exploits and Emergent Threats, LEET'08, pp 11:1–11:8, Berkeley, CA, USA, USENIX Assoc

38. Carlinet L, Me L, Debar H, Gourhant Y (2008) Analysis of computer infection risk factors based on customer network usage. In: Emerging Security Information, Systems and Technologies, SECURWARE Aug 2008. Second International Conference, pp 317–325

39. Lalonde L'evesque F, Nsiempba J, Fernandez JM, Chiasson S, Somayaji A (2013) A clinical study of risk factors related to malware infections. In Proceedings of the 2013 ACM SIGSAC Conference on Computer & Communications Security, CCS '13, pp 97–108, New York, NY, USA
40. Shin S, Lin R, Gu G (2011) Cross-analysis of botnet victims: New insights and implications. In: Sommer R, Balzarotti D, Maier G (eds) Recent Advances in Intrusion Detection, vol 6961 of Lecture Notes in Computer Science, Springer, Berlin Heidelberg, pp 242–261.
41. Huang DY, Dharmdasani H, Meiklejohn S, Dave V, Grier C, McCoy D, Savage S, Snoeren AC, Weaver N, Levchenko K (2014) Botcoin: Monetizing stolen cycles. In: Proceedings of the 2014 Network and Distributed System Security Symposium, San Diego, CA, USA

Chapter 3
Methodology and Measurement

3.1 Research Question

The principal goal of our study was to characterize the vulnerability of different countries to malware. Figure 3.1 shows the average attack frequency per host by country. Western Europe, the US and Australia have low attack counts, followed by Latin America, Russia, China and finally India has the highest attack frequency.

This study investigates how a range of variables including user-level factors, non-individual factors, and macro-economic factors explain the differences in attack frequencies we observed.

This chapter examines the data and metrics used to measure cyber-attacks on hosts in different countries.

3.2 Data

Our main data source is Symantec's Worldwide Intelligence Network Environment (WINE) [1]. Symantec's anti-virus products collect various types of information from users who opted-in to the data sharing program. We primarily analyzed two datasets: (1) WINE's binary reputation dataset which includes information on binaries that are present on participating hosts, and (2) WINE's anti-virus telemetry dataset which includes reports about host-based threats (e.g. viruses, worms, Trojans).

From the source data, we created two datasets: the 2010 dataset consists of all hosts active in the September 2009 to August 2010 period, the 2011 dataset consists of all hosts active in the September 2010 to August 2011 period. We analyze only attacks that occurred during these periods. However, to determine the state of the host (i.e. the binaries on a host), we analyzed all binary reports that had been submitted until the end of the respective period.

© Springer International Publishing Switzerland 2015
V.S. Subrahmanian et al., *The Global Cyber-Vulnerability Report*, Terrorism, Security, and Computation, DOI 10.1007/978-3-319-25760-0_3

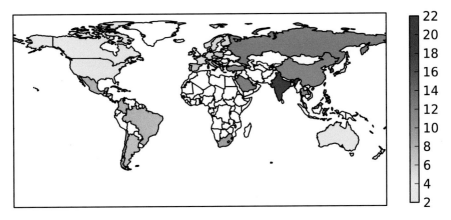

Fig. 3.1 Attack frequencies per host by country

Our main data source was supplemented with macro-economic data: the gross domestic product (GDP) per capita as published by World Bank [2] and the software piracy rate as published by the Business Software Alliance [3]. This data is discussed in greater detail below.

3.3 Dependent Variables

Vulnerability to cyber-attacks is measured by counting the number of malicious binaries identified on a host and aggregating the per-host data to the country-level. Before explaining how this was derived, it is important to clarify how malicious software (malware) is defined for this study and how the identification of a malicious binary is categorized.

As we discussed in Chap. 2, there are different types of malware. In most cases, their malicious nature is obvious. However, the categorization of binaries such as spyware is more complicated. Is software that has some useful functionality but also collects information about the user malware? At what point is a piece of software gathering enough information to be considered malware? This study relies on the judgment of the expertise of the malware analysts at Symantec. While there may be some debate about which software is classified as malware, there would be little influence on the global numbers in the WINE data set. More than 95% of the malicious binaries identified were either Trojans or viruses. The number of binaries in the "gray area" categories (such as spyware) is relatively low. As a consequence, we have high confidence that the counts in WINE of software deemed malicious are indeed correct.

Having defined what constitutes a malicious binary, it is also essential to discuss what information is revealed when such a file is found on a host. A malicious file can end up on a host in several ways. The binary could have been sent as an email attachment that was downloaded—usually automatically by an email program like

Windows Mail or Mozilla Thunderbird. The file could have been created by code that runs in memory after a successful drive-by attack by a malicious website. But the malware binary could also have been on an external medium like a USB thumb drive or a CD.

We regard all identified malware files as attacks with varying levels of success. A drive-by attack that was able to create a malicious file on a host successfully exploited a vulnerability in the web browser. Even if the attack was identified and blocked by anti-virus software and thus unable to establish permanence (i.e. create a binary and configure the system to run it at every start-up), the attack progressed far in the attack process and is counted as successful. In the case of malware that made it to a host via email, the attack is also counted because the malware may have been delivered to a mailbox whose provider does not scan for malware or could not identify the malware.

We use the term attack and not infection because many attacks are identified by anti-virus software immediately and never do any harm. But as any host that is not protected by anti-virus software would have been infected, we still see the attack as a serious threat.

E1: Average number of attacks per hosts

First, we use the mean of the number of attacks per host of a given country as a proxy for the vulnerability of that country's hosts:

$$\frac{\#\ \text{Attacks on all hosts of country c}}{\#\ \text{Host for country c}}$$

The plot in Fig. 3.2 shows the distribution of the number of attacks per host. Careful readers might wonder why we use the mean number of attacks per host and not the median number of attacks. Often the median is prefered to the mean as a more robust measure in the presence of outliers. But as the number of attacks per host has a long tail distribution the mean is the more informative measure. The median would hide the information in the tail of the distribution.

E2: Fraction of attacked hosts

Second, we computed the fraction of all hosts that encountered at least one attack:

$$\frac{\#\ \text{Host from country c with at least 1 attack}}{\#\ \text{Host for country c}}$$

The discussion of malware distribution (Sect. 2.1) indicates that there might be a non-linear connection between variables E1 and E2. One unpatched vulnerability could be the gate through which many pieces of malware enter a host. Likewise one successfully installed downloader on a compromised host could install an arbitrary number of other malware programs. Consequently the number of pieces of malware we find on a host might not accurately reflect the exposure to threats if one success-ful attack causes the installation of any number of malware pieces. Let's assume we have two identical hosts that have the same degree of vulnerability and caught the same malware downloader. If this downloader downloads two malware binaries to the one host and five malware binaries to the other host, should vulnerability be judged as equal?

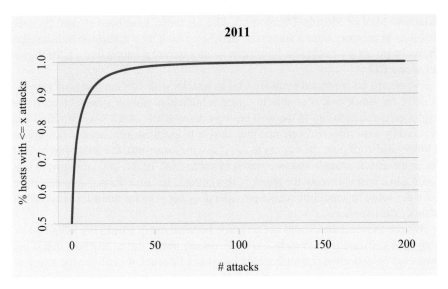

Fig. 3.2 Cumulative distribution plot of the number of attacks per host (for all hosts of the 44 countries analyzed)

In practice this problem is irrelevant. Figure 3.3 shows that the average number of attacks per host and the fraction of attacked hosts in a country are highly correlated (Pearson correlation is 0.89 for 2010 and 2011, Spearman rank correlation 0.95 and 0.97 for 2010 and 2011). A linear regression model $y=f(x)=z+a*x+b*x^2$ for the avg. attack number y based on the fraction x of attacked hosts with $z=14$, $a=-50$, $b=60$ for 2010 and $z=17$, $a=-63$, $b=72$ for 2011 explains the situation very well (adjusted R^2 of 0.85 and 0.88, respectively). Even if the attack frequencies do not always accurately reflect the exposure to threats at the level of individual hosts, this effect has been leveled out through the aggregation of data to the country level.

3.4 Features

In a previous study [4], we identified factors that are connected to increased probability of malware attacks on the user level.

3.4.1 Host-Based Features

H1: **Number of Binaries**
  For this feature, we count the total number of distinct binaries reported by a host. A binary file is any file that contains code. For machines running Microsoft Windows, this consists of all files in Portable Executable (PE) file format, including

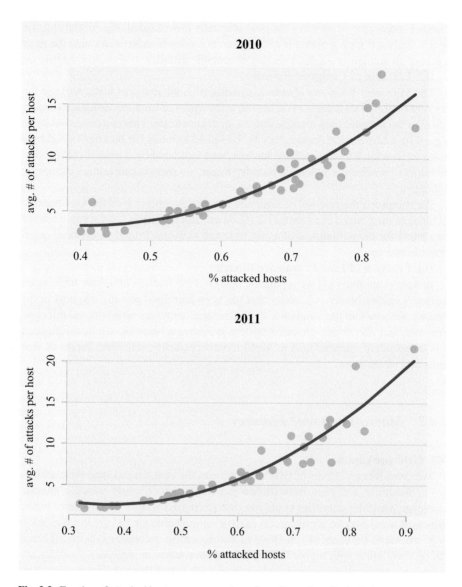

Fig. 3.3 Fraction of attacked hosts vs. avg. number of attacks per host for both datasets

executable files (.exe) and shared library files (.dll). Every time a binary gets updated—even if just a single bit is changed—a new binary is created and counted.

H2: Fraction of Downloaded Binaries

We count the number of binaries downloaded by the user from Internet sources. However, only user-initiated downloads are counted. Many more binaries are automatically downloaded by software components like Windows Update that update already installed software. The WINE environment records information on the

Internet source only if a binary has been manually downloaded. But for the purpose of this study it does not matter if a binary has been downloaded as a single file or as a part of an archive or a self-extracting executable installer.

H3: **Fraction of Unsigned Binaries**

The Microsoft Windows operating systems (that all analyzed hosts run) use the concept of digital signatures to ensure the authenticity of binaries. Software vendors can sign the binaries they create using a digital certificate. This certificate has to be signed by a trusted certificate authority. Unsigned binaries (or binaries signed by a software vendor whose certificate has not been validated by a trusted certification authority) constitute a potential security threat, as their manufacturer cannot be validated.

We computed the fraction of unsigned binaries as a portion of the total number of binaries on a host. Like feature H2 (% downloaded binaries), we see this feature as a proxy for the willingness of a user to install software from unknown and questionable sources.

H4: **Fraction of Low-Prevalence Binaries**

In a previous study [4] we separately analyzed the fraction of unique binaries on a host. A unique binary is a binary that has been identified on only one host in the dataset. We saw that the number of unique binaries provides no additional information over and above that provided by low-prevalence binaries. We therefore omit this independent variable here to avoid to unnecessarily increasing the set of features that we analyze.

3.4.2 Macro-Economic Features

M1: **GDP per capita**

We consider two effects of GDP per capita on the attack frequency. First, attackers in wealthier countries have higher incentives, which could increase attacks. However wealthier countries could also see decreased attacks on consumer hosts. Educated users may be more careful in their online behavior and greater education (e.g. measured by years of schooling or enrollment in secondary education) has a strong correlation with higher GDP [5]. Moreover, users in wealthier countries are better able to purchase devices and services with better security features (such as wireless routers with built-in firewalls or e-mail provider with automatic virus checking).

Another measure that parallels GDP per capita in its implications is Internet penetration rate (e.g. defined either as the percentage of households with an internet connection at home or defined as the percentage of citizens using the internet at least weekly). More sophisticated measures for the technological capabilities of countries have been proposed. The World Economic Forum's Global Competitiveness Report contains a technological readiness score [6], which is a compound measure with 50% of the score coming from ICT usage (% individuals using Internet, fixed and mobile internet subscriptions etc.).

M2: **Piracy rate**

Software from untrustworthy sources, like pirated software offered for free on the web or through file sharing services, comes with the risk of malware. Besides the direct impact of this variable through infected counterfeit software, it is reasonable to assume that people who get software from illegal sources also engage in other risky behavior online.

This study uses the piracy rates published by the industry advocacy group Business Software Alliance [3]. BSA defines the piracy rate as the fraction of the total installed software units that are unlicensed. We regard these numbers as correct although we are not able to independent verify their accuracy.

3.5   Data Preparation

The WINE environment contains data from 4,237,955 hosts for the 2010 period and from 4,146,548 hosts for the 2011 period. For all hosts we computed scores that we later aggregated by country to derive the values of independent host-based features and the dependent variables. Before we could aggregate the data, we had to clean it.

First, we removed all hosts that were active for less than 100 days in the respective period. We defined the activity period as the time difference between the first and the last submitted binary report. If a host was active for only a short time, there was only a limited potential for attacks. There are many reasons why a host may have short period of activity. The hosts might have installed the Symantec anti-virus software late in our observation period, or the user might have removed the software, or revoked the opt-in to the data-sharing program.

After this step we were left with 2,483,865 hosts for 2010 and 3,110,089 hosts for 2011. To remove further outliers we removed all hosts with less than 1500 or more than 15,000 binaries. We also removed all hosts with more than 200 attacks. This approach was taken to remove hosts that have not been actively used (due to a low number of binaries) or where data collection problems might have occurred (e.g. duplicate random host identifiers). Cleaning data like this is a crucial step in analyzing big data. We counted more than 50,000 malware reports for some host ids. These data collection errors would distort the results.

Finally we were left with 1,136,322 hosts for 2010 and for 2011 2,622,899 hosts. The data of these hosts was aggregated to the country level. The 44 countries that had at least 500 hosts in the cleaned datasets for both time periods are analyzed in the next sections.

3.6   Data Analysis

In this study we analyze attack frequencies as a proxy for the vulnerability to attacks.

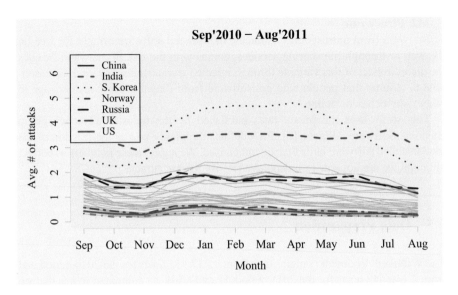

Fig. 3.4 2011 Monthly attack frequencies per host

3.6.1 Time-Series Analysis of Attacks

Figure 3.4 shows similar patterns for attack frequency across for most of the countries included in this survey. Most of the countries in the survey average less than two attacks per host per month per host. India and South Korea are the most significant outliers, averaging three or more attacks per host on most months. South Korea also has the strongest monthly variations in attack frequency; in most of the other countries attack frequency is fairly consistent from month to month.

The time-series chart in Fig. 3.5 also shows that the average number of attacks per host for a country is fairly stable. With respect to the reference month September 2010, the scores stayed in the +/−50% band for most countries. This stability highlights how historic attack counts are a good predictor for future attack frequencies. This result was expected, given that underlying factors like human risk behavior, education and technology capabilities change slowly.

3.6.2 Macro Economic Factors

Table 3.1 shows strong and consistent correlations between macro-economic factors such as GDP, piracy rate, and internet penetration. GDP and piracy rate are both strongly positively correlated with the log average number of attacks and attacks per host, while Internet penetration is strongly negatively correlated with the log average number of attacks. As will be discussed in greater detail below, overall nations with higher per capita GDP are less likely to be attacked, while higher software piracy rates are correlated with more frequent attacks.

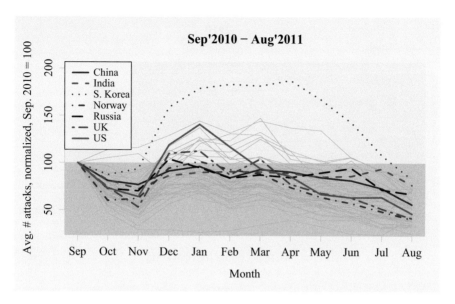

Fig. 3.5 2011 Monthly attack frequencies normalized by September 2010 values

Table 3.1 Pearson correlation coefficient between macro economic factors and the log of the average number of attacks per host by country for the 2010 and 2011 datasets

Feature	Pearson 2010	Correlation 2011
GDP	0.81	0.82
Piracy rate	0.65	0.70
Internet penetration	−0.74	−0.75

3.6.2.1 GDP per Capita and Attacks

Figure 3.6 shows the relationship between GDP per capita and the average number of attacks on the hosts of a country. Although the financial incentive to attack the wealthiest countries is presumably higher, there are fewer attacks on U.S. and European hosts.

More detailed results providing breakdowns according to specific attack types are shown in Fig. 3.7, overall they conform to the overall pattern.

3.6.2.2 Piracy and Attacks

Figure 3.8 shows a clear statistical correlation between attacks and the piracy rate (Pearson correlation=0.70). However, piracy is also strongly negatively correlated with GDP per capita.

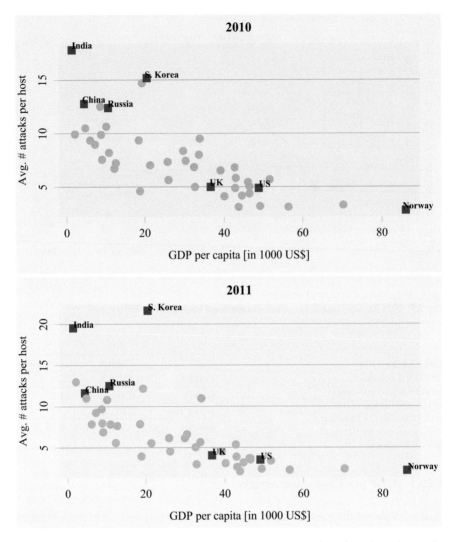

Fig. 3.6 Relationship between the piracy rate and the average number of attacks on hosts of a country in 2010 (*top*) and 2011 (*bottom*). Selected countries highlighted

Figure 3.9 shows detailed results for six malware types. For viruses, Trojans, and worms, which represent the vast majority of attacks, the correlation between increased software piracy and increased attacks is consistent. For less frequent attacks, particularly misleading software and adware, the relationship between increased piracy and increased attacks is not as clear. However, these attacks represent only a small portion of the total.

However, our multivariate linear regression model suggests that the correlation between piracy rate and attacks is inconclusive. When the model includes GDP, the piracy rate does not matter.

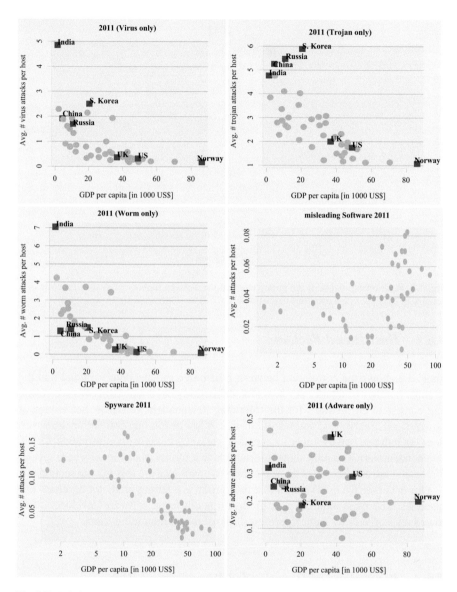

Fig. 3.7 Relationship between the GDP per capita and the average number of attacks on hosts of a country for virus, Trojan, worm, misleading software, spyware and adware attacks. Selected countries highlighted

The piracy rate is probably a less useful measure for our purposes because it includes all types of unlicensed software and does not distinguish between different sources. For example, the overall piracy rate does not distinguish between an installation from a genuine copy of the software (i.e. a CD provided by the software manufacturer) installed on more than the licensed number of machines or from a pirated CD bought from a street dealer.

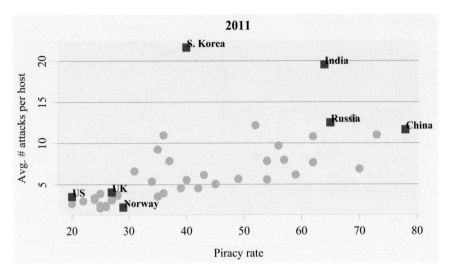

Fig. 3.8 Relationship between the piracy rate and the average number of attacks on hosts of a country in 2011. Selected countries highlighted

3.6.3   Host-Based Factors

Table 3.2 shows the correlation between host-based independent variables and the average number of attacks per hosts for the periods 2010 and 2011. We see strong and consistent correlations for the following independent variables: percentage of downloaded binaries and number of binaries. A low or missing correlation, however, does not indicate that the other independent variables contain no information. Often, independent variables only show information in a multivariate model where several dimension are considered at the same time. We will analyze this in later sections.

Our data (see Chap. 4) shows that user group is a strong predictor of attack frequencies. The most attacked user group turns out to be software developers, followed respectively by gamers, professional users and remaining users.

3.6.4   Percentage of Downloaded Binaries and Risk

Figure 3.10 shows a strong linear relationship between the percentage of downloaded binaries and the number of attacks per host. Though a few countries are exceptions to this trend (e.g. China), the trend appears to hold for most countries.

We have already seen that per capita GDP explains much of the differences in attack frequencies. The residuals, i.e. the differences between the predicted attack frequencies according to the regression model and the observations represent the unexplained variability. Figure 3.11 shows the relationship between the fraction of downloaded binaries and these residuals.

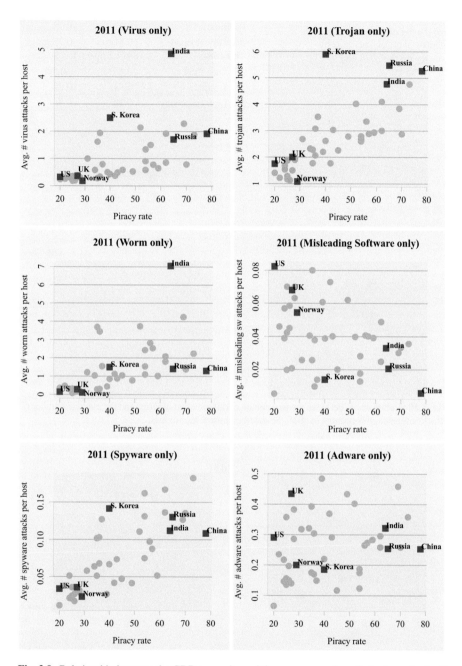

Fig. 3.9 Relationship between the GDP per capita and the average number of attacks on hosts of a country separately for virus, Trojan, worm, misleading software, spyware and adware attacks. Selected countries highlighted

Table 3.2 Pearson correlation coefficient between host based factors and the log of the average number of attacks per host by country for the 2010 and 2011 datasets		Pearson	Correlation
	Feature	2010	2011
	# Binaries	0.37	0.78
	% Downloaded binaries	0.51	0.73
	% Unsigned	0.15	−0.20
	% Unique	0.44	0.20
	% Low frequency	0.30	0.36

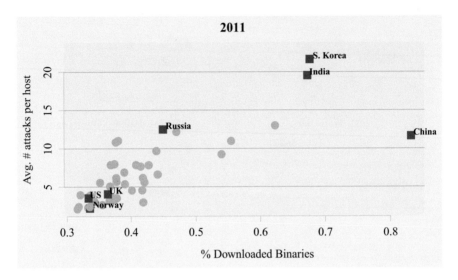

Fig. 3.10 Relationship between the percentage of downloaded binaries and the average number of attacks on hosts of a country in 2010. Selected countries highlighted

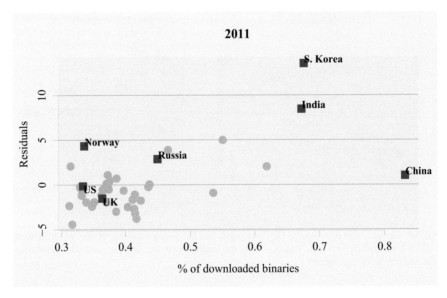

Fig. 3.11 Relationship between the percentage of downloaded binaries and the residuals of a linear model fitted by GDP. Selected countries highlighted

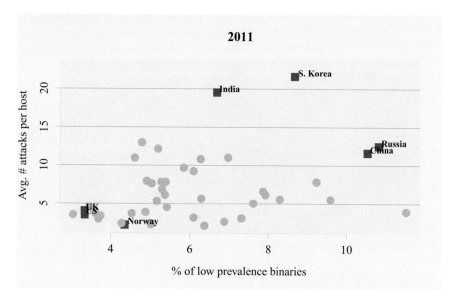

Fig. 3.12 Relationship between the percentage of low-prevalence binaries and the average number of attacks on hosts of a country in 2011. Selected countries highlighted

3.6.5   *Percentage Low Prevalence Binary and Risk*

Figure 3.12 also shows an increasing linear relationship between the percentage of software on hosts that constitute low prevalence binaries and the average number of attacks per host.

3.6.6   *Percentage Unsigned Binaries and Risk*

Figure 3.13 shows an increasing linear relationship between the percentage of binaries on a host that are unsigned and the average number of attacks per host. Likewise, Fig. 3.14 shows an increasing linear relationship between the percentage of binaries on a host that are unsigned and the percentage of infected hosts in a country. Simply put, countries in which hosts tend to have lots of unsigned binaries tend to be increasingly vulnerable to cyber-attacks.

3.6.7   *Attacks of Misleading Applications and Spyware*

We analyzed our data separately for two subgroups of malware: misleading applications and spyware. Both misleading applications (like fake anti-virus software) and spyware (which logs key strokes to collect credit card data) are directly targeting

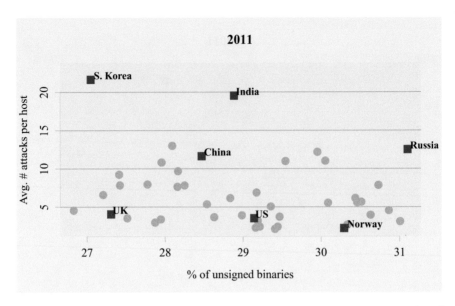

Fig. 3.13 Relationship between the percentage of unsigned binaries and the average number of attacks on hosts of a country in 2011. Selected countries highlighted

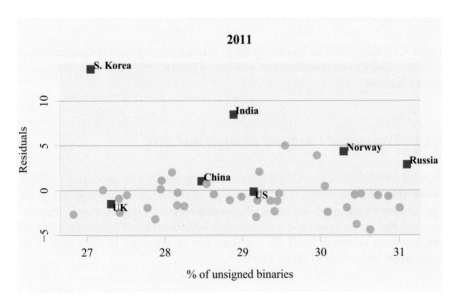

Fig. 3.14 Relationship between the percentage of unsigned binaries and the residuals of a linear model fitted by GDP. Selected countries highlighted

users. Vulnerability to those attacks might be different from vulnerability to malware in general, which is often indifferent towards the user and only exploits the
infected hosts' computing power and network connectivity.

To our surprise, spyware and misleading applications showed very different patterns in our data (Fig. 3.15). While spyware shows roughly the same pattern as
attacks in general (Fig. 3.6), we see a positive correlation between misleading soft-

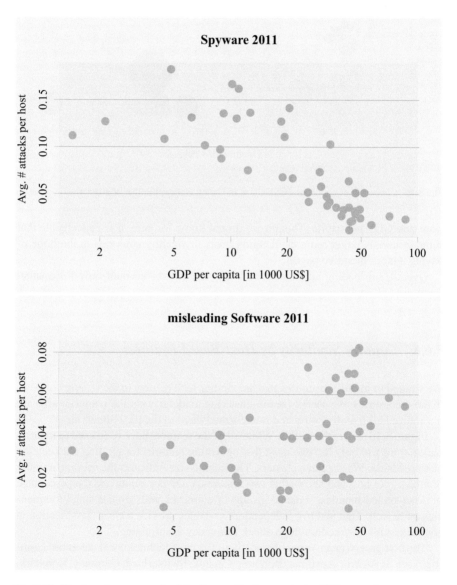

Fig. 3.15 Number of spyware and misleading application attacks by GDP per capita

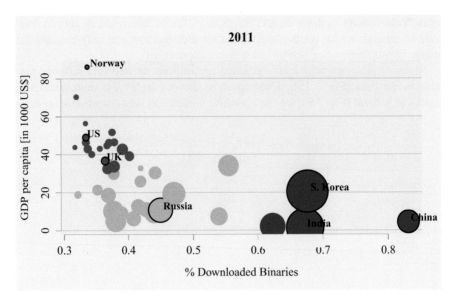

Fig. 3.16 Clustering of countries. Bubble size indicates average number of attacks

ware and GDP per capita. Though we do not know for sure, it is conceivable that some criminals target more challenging hosts in wealthy countries in the hope of having a more lucrative payoff.

Overall, misleading application and spyware attacks account only for a small fraction of all malware attacks.

3.6.8 Country Similarity by Host-Based Features

We wanted to identify countries that are similar with respect to the features of hosts in those countries. We used k-means clustering to identify similar countries after we normalized all feature values and transformed them to the [0, 1]-interval.

Figure 3.16 shows the results. While all of the features have been used to identify clusters, we use only the two most distinguishing features for plotting the keep the chart readable. We see three clusters. The bubble size indicates the average number of attacks per host and shows that two features, GDP per capita and the percentage of downloaded binaries, separate groups of countries with similar attack frequencies quite well. That just two independent variables provide a clear classification in countries with low/medium/high attack frequency is surprising.

The first group (green) consists mostly of the most technological advanced countries such as Austria, Australia, Belgium, Canada, Switzerland, Germany, Denmark, Hong Kong, and Singapore.

The second group includes middle-income nations such as Argentina, Brazil, Czech Republic, Greece, and Turkey.

The third group, which suffers the greatest number of attacks also has the lowest income levels overall and include China, India, South Korea, the Philippines and South Africa.

3.6.9   Multivariate Analysis

In the previous sections we discussed each independent variable separately. We already have shown that the GDP per capita has strong explanatory power and that other factors can explain some of the remaining variance in the data. We now provide a multivariate analysis that takes all variables into account at once. We are interested in identifying those factors that contribute most to the explanation of the observed attack frequencies while controlling for the other variables in the models.

We fitted a linear regression model. As our prior analysis already indicated, the attack frequencies show a saturation effect. Attack frequencies decrease with an increase of wealth and a decrease of risky behavior. But the marginal effect of the independent variables decreases, i.e. every additional dollar of GDP per capita decreases the attack risk slightly less than the previous dollar. So our model is $\log(y_i) = \Sigma_j a_j x_{ij}$ where y_i is the average number of attacks of country i, x_{ij} is the value of the j-th predictor variable of country i and a_j is the fitted parameter of the linear regression model.

The parameter estimates and their associated p-values are shown in Table 3.3. As can be seen from Fig. 3.17, the model fits the observed attack frequencies very well.

In Fig. 3.16 we saw that these two variables separate groups with low/medium/high attack counts pretty well.

Table 3.3 Parameter estimates and significance of regressors for the fitted Poisson regression model

Regressor	Estimate	p-Value	Sign. level
(Intercept)	1.954e+00	0.237820	
GDP per capita	−1.152e−05	0.018585	*
Piracy rate	5.335e−03	0.257633	
Avg. # binaries	−2.923e−04	0.136608	
Avg. % downloaded binaries	2.130e+02	0.000257	***
Avg. % unsigned binaries	2.803e+00	0.564242	
Avg. % low prevalence binaries	−1.453e+00	0.608494	

Significance levels: $*p<0.05$, $**p<0.01$, $***p<0.001$
Adjusted R^2: 0.7875, p-value: 1.044e−11

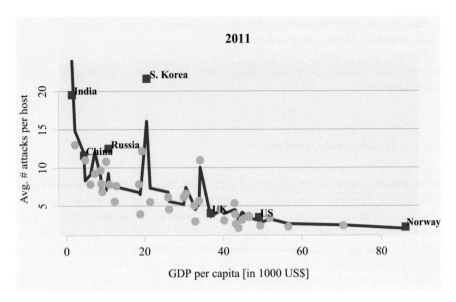

Fig. 3.17 Observations (*dots*) and expected values (*line*). Model created using all predictor variables

3.6.9.1 Model Verification and Interpretation

The estimated model fits the data well, but it is possible that we estimated a model that fits the data too well, i.e. that there is a risk of overfitting. To account for the potential for overfitting the model, we checked the predictive power of the model and conducted a month-to-month analysis. In other words, we estimated the model parameters for 2011 with the data from month t and predicted the average number of attacks in month $t+1$. As the baseline predictor, we use the average number of attacks in t. Using total residual standard error as a measure, the fitted model provides a lift of 2.3 compared to the baseline. That means that the model predicting attacks based on *GDP per capita* and *% downloaded binaries* decreased the error by over 50%.

Still, the model does not explain the total variation. Using a country's previous month's attack frequency, the total squared deviance between predicted and observed attacks could be decreased from 75 to 25. Because the purpose of this study is not to predict attack frequencies but rather to explain what factors correlate with attacks, this result is sufficient. In summary, we can conclude that *GDP per capita* and *% downloaded binaries* can be seen as the best measures for two underlying factors: the technological capabilities of a country and the average tendency of its citizens to expose themselves to malware threats. All of the other independent variables are highly correlated to one of these factors.

This is a reasonable result. A successful attack requires a malware distribution method. With insecure JavaScript practices, unsecured web servers, or search engines without proper filtering technology for malicious webpages, the technology environment determines how hard and costly it is to set up an attack. The personal risk-related behavior of users—keeping software updated or getting software only from reliable sources—determines the success rate of an attack.

But our model has some limitations. Most strikingly, the model cannot explain the high attack frequencies encountered by South Korea. Extreme outliers always raise suspicions of data errors. After double-checking for potential sources of errors, we noted that South Korea has a very unique threat situation which may explain its high attack frequency. Malware attacks targeted specifically to South Korea have been observed in the past [7]. Symantec [8] observed in 2012 (after our observation period) a threefold increase in phishing attacks on South Korea (where social engineering methods are used to make users provide fake websites with personal information like login credentials or credit card details). Attacks on South Korea are assumed to be partially politically motivated—a dimension not covered in our study.

References

1. Dumitras T, Shou D (2011) Toward a standard benchmark for computer security research: The Worldwide Intelligence Network Environment (WINE) Apr 2011. In: EuroSys BADGERS Workshop, Salzburg, Austria
2. World Bank. International comparison program database. http://data.worldbank.org/.
3. Business Software Alliance (2012). Shadow market—2011 BSA global software piracy study, 2012. 9th edition, http://globalstudy.bsa.org/2011/downloads/study_pdf/2011_BSA_Piracy_Study-Standard.pdf
4. Ovelgönne M, Dumitras T, Prakash BA, Subrahmanian VS, Wang B (2015) Understanding the Relationship between Human Behavior and Susceptibility to Cyber-Attacks: A Data-Driven Approach, under revision for a technical journal, Feb 2015
5. Sianesi B, Van Reenen J (2003) The returns to education: Macroeconomics. In: J of Economic Surveys, 17(2):157–200
6. The global competitiveness report 2010–2011 (2010) Technical report, World Economic Forum
7. South Korean malware attack (2013) Technical report, United States Computer Emergency Readiness Team (US-CERT), https://www.us-cert.gov/sites/default/files/publications/South%20Korean%20Malware%20Attack_1.pdf
8. Symantec Corporation (2013) Symantec Internet security threat report, vol 18 http://www.symantec.com/content/en/us/enterprise/other_resources/b-istr_main_report_2011_21239364.en-us.pdf

Chapter 4
Human Behavior and Susceptibility to Cyber-Attacks

For each of the 35 possible combinations (five categories times seven independent variables), we studied the relationship between each of these seven independent variables and one dependent variable, namely the number of attempted malware attacks detected by Symantec on the machine. Our results show that the first variable (the number of binaries on a machine) is closely linked to number of attacks for software developers, while the next five are linked to the number of attacks for all user categories. Surprisingly, our results show that software developers are more at risk of engaging in risky cyber-behavior than other categories.

4.1 Introduction

Cybersecurity systems based on rigorous theoretical proofs often fail in practice— many researchers consider human users to be the weakest link in the system [1–3]. The extent to which user behavior is related to the propensity of a host to be the target of cyber attacks is not well understood. In ordinary crimes such as muggings, criminals often select their victims on the basis of their behavior and characteristics (e.g. walking late at night, being old and infirm, etc.). We expect that the likelihood and intensity of cyber attacks against an end-host are influenced by user behaviors such as downloading and executing software of uncertain provenance, connecting to multiple networks while traveling, or employing the computer for gaming. Some security experts warn that in the cyber-crime world "only amateurs attack machines; professionals target people" [4].

Humans can fall prey to social engineering attacks which request them to visit Web sites or to download files that result in the installation of malware; they may also visit compromised Web sites that conduct drive-by-download attacks [5] which exploit vulnerabilities in browsers causing silent file downloads. In short, users often willingly download unknown applications and binaries or unknowingly perform actions that undermine security.

© Springer International Publishing Switzerland 2015

V.S. Subrahmanian et al., *The Global Cyber-Vulnerability Report*, Terrorism, Security, and Computation, DOI 10.1007/978-3-319-25760-0_4

This chapter focuses on the *problem of identifying human behaviors which increase the risk of malware attacks on a host*. We systematically analyze the user behaviors and cyber attacks observed between January–August 2011 on 3.5 million end-hosts. The data is available through the Worldwide Intelligence Network Environment (WINE) [6] described earlier in the book. WINE consists of several data sets of which the following two are used in the work reported in this chapter:

- WINE's *binary reputation* data set includes information on binary executables downloaded by users who opt-in for Symantec's reputation-based security program.
- WINE's *anti-virus telemetry* data set includes reports about host-based threats (e.g., viruses, worms, Trojans) detected by Symantec's anti-virus products. As mentioned in previous chapters, because this data is collected on hosts targeted by cyber attacks—rather than honeypots or small-scale lab settings—it provides a unique window into the factors affecting security of real computer users worldwide.

We identify several features that point to specific human behaviors, and we analyze how the risk of cyber attacks changes with different behaviors. As WINE contains no information that would allow us to identify users, we assume that each host in our data corresponds to one user and we assess the user's behavior anonymously using the events recorded on that host. We estimate the risk of attacks using the frequency of malware detections on each host. As all the hosts in our dataset were protected by Symantec products, the observed attacks were actually *blocked*, but had the machine not had an anti-virus program installed, many attacks would have succeeded. Thus, our measurement of risk is actually directed at the population of machines not protected by any anti-virus program.

Table 4.1 summarizes our key findings showing that for all categories of users, the number of low-prevalence binaries downloaded by the users, number of unique binaries on users' machines, number of unsigned binaries on the users' machines, and number of binaries downloaded by users are all associated with an increase in the number of malware attacks. In the case of software developers, the number of

Table 4.1 Overview of analyzed independent variables

Ind. variable	Gamer	Pro	SW-dev	Other	All
# Binaries	✱	✱	✓	✱	✱
% Low-freq bin.	✓	✓	✓	✓	✓
% Hi-freq bin.	✓	✓	✓	✓	✓
% Unique bin.	✓	✓	✓	✓	✓
% Unsigned bin.	✓	✓	✓	✓	✓
% Downloaded bin.	✓	✓	✓	✓	✓
# of ISPs	✱	✱	✱	✱	✱

✓ Mark shows a relationship between the independent variable and the number of attempted malware attacks
✱ Mark shows that we don't see enough evidence for a relationship

binaries they installed on their machine is also related to the number of attacks. Finally, we note that the number of ISPs used to connect to the network has a statistically significant influence on the number of attacks. However, we think the magnitude of that influence is too low to claim a relationship given the potential sources of error we discuss later.

Implications for the security industry. It is increasingly difficult to protect users against malware because of the growth in volume and diversity of cyber-attacks. Characterizing the user behaviors that are more likely to attract cyber-attacks opens up new opportunities for identifying and defending the hosts that are at risk. For example, security analysts estimate that 403 million new malware samples were created in 2011 [7]. This growth results in a large number of low-prevalence files, which are present on few hosts and are likely to be malicious, as attackers employ polymorphism techniques in order to evade detection. This observation represents the basis of recent *reputation-based security* techniques [8, 9], which compute a reputation score for each unknown file based on features such as the file's prevalence in the wild, before analyzing the content of the file. Today, reputation-based security systems are included in several anti-virus products, as well as in the Windows 8 operating system [10]; however the association between the users' propensity to download low-prevalence files and cyber attacks has not been validated at a large scale.

Similarly, best security practices recommend reducing the *attack surfaces* of end-hosts [11]. Attack surface reduction works by decreasing the number and severity of potential attack vectors that each host exposes (e.g., open sockets, RPC endpoints, running services). Even if the software contains vulnerabilities—perhaps not yet discovered—the attacks will succeed only if a corresponding attack vector is available. However, users can alter the attack surfaces of their computers by downloading and installing new software, which may enable additional attack vectors. By helping analysts understand how the number of binary executables present on a host affects the volume of cyber attacks, our work allows them to assess the impact of attack-surface reduction techniques on security in the field.

Roadmap. This chapter is organized as follows: we first describe our dataset, and then the statistical features signifying human behaviors of interest. We then present our approach and hypotheses. Finally we review related work and conclude with implications of our findings.

4.2 Dataset and Set-up

To characterize the link between human behavior and cyber attacks, we integrated information in several datasets collected from different observation perspectives. We describe our problem statement, and the datasets we used (including associated caveats) for our research in this section.

4.2.1   Problem Statement

More specifically, our research problem can be defined as follows:

GIVEN: Security telemetry from Symantec's WINE datasets (details in Sect. 4.2.2).

FIND: The statistical proxies of human behaviors which are related to increased malware reports on a machine.

Clearly, addressing this problem involves (a) extracting carefully constructed data-based features from the WINE datasets, and then (b) performing sound statistical tests to relate the *independent* behavioral variables (the features) to the *dependent* variable (the number of malware detections).

4.2.2   The WINE Datasets

As mentioned in previous chapters, Symantec's WINE data is collected from real-world hosts running their consumer anti-virus software. Users of Symantec's consumer product line have a choice of opting-in to report telemetry about the security events (e.g. executable file downloads, virus detections) that occur on their hosts. The events included in WINE are representative of events that Symantec observes around the world [12]. WINE enables reproducible experimental results by archiving the reference data sets that researchers use and by recording information on the data collection process and on the experimental procedures employed.

We analyze the complete set of events recorded in the *binary reputation* and *anti-virus telemetry* data sets from WINE during the 8 month period between January–August 2011. For this period, the dataset contains 13.7 billion reports collected on 3.5 million hosts. WINE does not include user identifiable information.

Anti-virus telemetry. Anti-virus (AV) telemetry records detections of known malware for which Symantec generated a signature that was deployed in an anti-virus product. As commercial security products generally aim for low false-positive rates, we have a high degree of confidence that the files detected in this manner are indeed malicious.

From each record, we use the detection time, the associated threat label, the hash (MD5 and SHA2) of the malicious file detected and the manner of the detection (signature scanning or behavioral features extracted from an execution of the file on the end host). Each record indicates that the anti-virus has blocked an attack that may have resulted in an infection.

Binary reputation. The binary reputation data records all binary executables—benign or malicious—that were downloaded/copied on end-hosts worldwide. From each record we extract the time stamp of the file creation event, the country in which the host is located, the hash (MD5 and SHA2) of the binary, and the URL from which it was downloaded (if available).

Data cleaning. First, we restrict our analysis to the 20 countries with the most hosts in the dataset. That left 2.9 million machines. We also removed any machine that was active for less than 200 days in the 8-month period. We take the range of activity as a proxy for potential virus threats. After applying the 200 day filter rule, 1.7M machines remained. The percentage of machines that did send reports for the full period studied seems to be very high. However, it seems very likely that either many hosts revoked the data sharing opt-in at some point or replaced the Norton software with something else (or nothing) after their license expired. Some hardware vendors sell computers with pre-installed trial versions of Symantec's software. Conversion rates from trial users to paying customers are usually low for any type of product or service. We removed outliers (hosts that have more binaries or attacks that are more than two standard deviations away from the mean) we ended up with a *cleaned dataset* of 1.6M hosts—all our results were derived from this cleaned data.

Figure 4.1 shows the distribution of number of malware found per host. Encouragingly, more than 50% of hosts in the cleaned data encountered no malware during our observation period. Most machines had 50 attacks or less.

Shortcomings of our study. As WINE does not include telemetry from hosts without Symantec's anti-virus products, our results may not be representative of the general population of platforms in the world. In particular, users who install anti-virus software might be more careful with the security of their computers and, therefore, might be less exposed to attacks. Additionally, our data was collected on hosts running various versions of Windows; the trends we observe may not apply to other operating systems. Although we cannot rule out the possibility of such selection bias, the large size of the population in our study (3.5 million hosts, or 1.7M

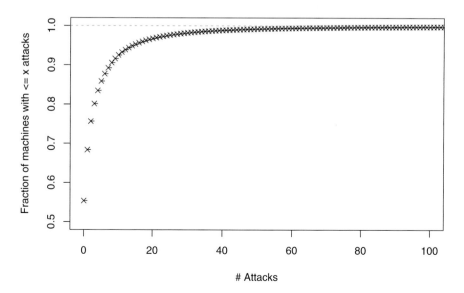

Fig. 4.1 Empirical cumulative density of the number of malware attacks per machine

hosts after down-selection as described above) and the fact that Windows has been the primary target for cyber attacks for the past decade suggest that our results have a broad applicability. The anti-virus applications that gather WINE data operate on end-hosts. Hence, we do not know how many attacks are deflected by security measures in the environment (e.g., a hardware firewall, or intrusion-prevention services provided by an Internet Service Provider) or by operating system defenses that sit in front of Symantec's software. Therefore, the number of attacks observed in the anti-virus telemetry data should be interpreted as a lower bound.

Finally, the WINE datasets do not provide sufficient information for determining the rate of successful attacks on the targeted hosts. However, by correlating the attacks blocked and the files present on each host in our data set, we believe we can derive unique insights on the link between human behavior and susceptibility to cyber attacks.

4.3 Feature Construction

The goal of this chapter is to analyze human behaviors w.r.t. cybersecurity. Hence an important step is constructing statistical features based on the WINE datasets, which can act as proxies for human behaviors. To this end, we analyzed the following features (i.e. the 'independent variables'):

1. **Number of binaries present on a host**

    DEFINITION: The number of executable files on a machine.
    MOTIVATION: The total number of executables represents a measure of the host's attack surface because each executable file may include known or unpatched vulnerabilities and may provide distinct attack vectors.
    ASSOCIATED BEHAVIOR: Indiscriminate installation of software. The human user can increase this attack surface by downloading/installing more executables.

2. **Percentage of low-prevalence binaries on a machine**

    DEFINITION: A low-prevalence binary is one that is present in less than 1000 hosts in our cleaned WINE dataset. The percentage of low-prevalence binaries is the ratio of number of low-prevalence binaries to number of binaries in the cleaned WINE dataset.
    MOTIVATION: Hackers often create numerous minor polymorphic variants of malware in order to evade detection—these are low-prevalence files. This is one of the key observations behind modern reputation-based security technologies [8, 9].
    ASSOCIATED BEHAVIOR: Some users have a tendency to accumulate low prevalence binaries (e.g. by downloading "cheats" in online games which are often infected with malware [13] or by downloading free software).

3. **Percentage of high-prevalence binaries on a machine**

DEFINITION: A high-prevalence file is one that is present in over 1M hosts in our cleaned WINE dataset.

MOTIVATION: We studied high-prevalence files solely in order to complement our study of low-prevalence files.

ASSOCIATED BEHAVIOR: Some users have a tendency to download popular binaries, perhaps binaries that lots of their friends are downloading such as new social network apps.

4. **Percentage of unique binaries**

DEFINITION: A binary that appears on only one host represents an extreme case of low prevalence.

MOTIVATION: As discussed above, a unique binary may be a polymorphic variant of a piece of malware.

ASSOCIATED BEHAVIOR: A user who downloads unique binaries is one who is downloading low-prevalence files as mentioned above. We note, however, that software vendors sometimes create unique binaries by embedding digital watermarks or customer-specific information, for licensing purposes and some unique binaries may fall into this category.

5. **Percentage of unsigned binaries on a machine**

DEFINITION: Commercial software vendors usually sign their binaries digitally to verify the integrity of software and to establish the identity of the vendor. This feature measures the percentage of binaries on a host that is unsigned. Lack of a digital signature does not necessarily mean that the binary is malicious; for example, open-source software is typically distributed without being signed.

MOTIVATION: As unsigned binaries do not have a reputed entity affirming the integrity of the binary, one may hypothesize that these binaries are more likely to be malware.

ASSOCIATED BEHAVIOR: A user with a high percentage of unsigned binaries exhibits a tendency to "go with" fewer well known software vendors. This may indicate that he cares less about the reputations of the vendors whose software he installs on his machines.

6. **Percentage of downloaded binaries**

DEFINITION: Percentage of binaries that the user downloads from the web, intentionally or otherwise.

MOTIVATION: Malware is often distributed via the Web [5].

ASSOCIATED BEHAVIOR: Users who download a high percentage of binaries from the web (rather than receiving them as email attachments or copying them from a physical medium, such as a CD-ROM or USB drive) may be visiting more questionable sites, especially if these downloaded binaries are not signed.

7. **Travel history of a user**

DEFINITION: This is the number of ISPs from which a host has connected to
the network.
MOTIVATION: In a time when many users use laptops, tablets, and smartphones
for their computing needs, there is a high probability that these machines
"travel". People carry laptops from home to work to conference sites, air-
ports, and hotels. At each such site, their machine may connect to a local, less
secure, ISP.
ASSOCIATED BEHAVIOR: Individuals who feel an absolute need to be connected through free Wi-Fi networks may value connectivity more than security. By using the number of ISPs that a machine connects to as a proxy for a
user's travel habits, we wondered whether amount of travel by a user can be
linked to the risk of malware attacks on the machine.

4.4   User Classification

Cyber-attackers may target different categories of users by exploiting vulnerabilities in software primarily used for professional tasks (e.g. SAP) or by taking advantage of the fact that some users download more binaries than others (e.g. for gaming).
We classify users into four anonymous categories, based on the application programs present on their computers. As our curated WINE dataset includes 352.8
million binaries, we use the manufacturer information provided by the certificates
used to digitally sign binaries. We identify a total of 902 software development
companies that had signed at least 100K reported binaries. We then classify all users
via four categories: gamer, pro, SW-dev, other describing gamers, professionals
(other than software developers), software developers, and all others, as follows:

Pros. A host with more than 50 binaries from companies that only manufacture
professional software is considered a professional host.

Gamers. A host with more than 50 binaries from companies that only manufacture games is considered a gamer.

Software Developers. We created a list of popular tools (Table 4.2) and classified
any machine having any of these tools as a software development host.

Other. All hosts that have not been classified in one or more of the aforemen-
tioned categories are considered to be in the "other" category.

We note that our classification criteria may classify some hosts into more than
one category. We chose this approach because in real life, users may utilize one
computer for multiple purposes, e.g. a computer science student who completes
programming assignments and plays games on the same laptop.

Figure 4.2 shows the average values of the features described in Sect. 4.3 by user
category. We normalize these values by the averages for all users in each category,
to highlight the behaviors that are typically associated with each user category. The
most striking deviation from the average values of all machines are the high fraction

Table 4.2 Definition and examples for classification of software vendors and software

Category	Description	Examples
Professional	Software vendors whose products are only used in professional contexts and have no dual consumer/business use like office packages, examples are vendors for ERP, CAD, data center or call center software	SAP, EMC, Sage Software, Autodesk, Dassault Systemes, Citrix, TiFiC AB
Gaming	Every software vendor that publishes only gaming software including more traditional video games and/or social entertainment and virtual worlds, no multi-product companies like Microsoft	Valve, Electronic Arts, Blizzard, Duowan, Epic Games, WildTangent, Jorudan, IMVU
Software development	Software Development Software used in the software development process including compilers, IDEs, version control systems	VisualStudio, Eclipse, NetBeans, Java SDK, Subversion, Git, Mercurial

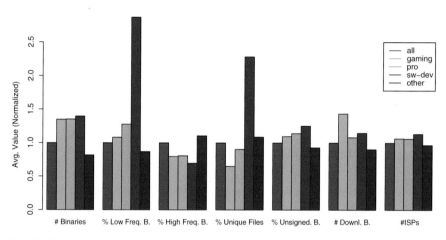

Fig. 4.2 Independent variable averages by user category (normalized, values for all set to 1)

of low-prevalence and unique files of software developers. This is not surprising. When a software developer compiles code, a new and most likely unique binary is created. Additionally, Fig. 4.2 shows that gamers have a higher than normal number of downloaded binaries on their computers and SW-devs spend an above average fraction of time working at night.

Observation 1. *On average, gamers encountered 83% more malware attacks than non-gamers, while professional users encountered have 33% more malware attacks than non-professional users.* As gamers have the habit of downloading more binaries from the Web than other users, this puts them at risk. Malware that targets popular gaming platforms, e.g. with the aim of stealing World of Warcraft credentials [14], further raises the risk profile of gamers. The higher risk of cyber attacks against professional users reflects the recent increase in targeted attacks aimed at stealing sensitive/proprietary information from corporations [15, 16]. As the ability to remain stealthy (e.g. through the use of zero-day exploits or advanced persistent threats) is key in these attacks, they typically target only a few selected employees—hence, such attacks do not result in a large volume of malware aimed at professional users.

Observation 2. *Interestingly, we observe the biggest difference in the volume of malware for software developers: they have 145% more malware than non-developers*. On average, we found significantly more malware (8.1) on SW-dev hosts than on non-SW-dev hosts (3.3). Perhaps some software developers build tools to analyze malware, or they learn how to develop exploits, or they participate in vulnerability rewards programs. However, it seems more likely that, as the group of users most intimately familiar with the inner workings of computer systems, developers find ways around the restrictions imposed by firewalls and anti-virus software, e.g. in order to deliver a project on time, and they also tend to disregard security best practices that are aimed at regular users, by downloading risky binaries of unknown provenance. Perhaps software developers sometimes feel that their knowledge of software arms them with better protection against cyber attacks when compared with automated tools or simple rules of thumb. As the WINE data reflects attacks rather than infections (see Sect. 4.2.2), we do not know whether this is indeed true, or, conversely, whether their knowledge lulls software developers into a false sense of security. However, the WINE data shows that software developers attract considerably more cyber attacks than other users.

4.5 User Behavior and Cyber-Attacks

We analyze the relationship between each independent variable (IV; the features of Sect. 4.3) and the number of observed attacks. First, we fitted a quasi-Poisson regression model with a sqrt-link function to measure the validity of the connections between attacks and IVs. When we fit models separately for all seven IVs, we get p-values below 2×10^{-16} for all independent variables. This indicates that each of these features is statistically significantly linked to the number of attacks a host receives. In a multivariate model with all seven features, we get the following p-values:

Ind. variable	p-Value	Sig. level
# Binaries	0.0479	*
% Downloaded bin.	<2e−16	***
# of ISPs		
<2e−16	***	
% Unsigned bin.	0.8344	
% Low-freq bin.	<2e−16	***
% High-freq bin.	<2e−16	***
% Unique bin.	0.4163	

This indicates that the features for unique and unsigned binaries do not provide additional information when used in combination with the other features.

The insight of this first analysis is limited. Regardless of how goodness-of-fit could be measured (there is no standard approach for Poisson models), due to the high randomness in the malware attacks the fit and predictive power of a regression model is very low i.e. it is not useful for predicting the number of attacks on a specific machine.

However, if we aggregate the data and group hosts by their feature values, we see very clear trends which we will discuss in the remainder of this section. Also as we see from our later analyses, because our data contains information from 1.7M million hosts, it is highly unlikely that any observed pattern is random. This is further backed up by t-tests showing statistical significance at the 99.9% level (i.e. $p < 0.001$) as we will see in the rest of this chapter.

4.5.1   Analysis Methods

We analyze the behavioral features in three ways that are better suited to test our hypotheses. For a given IV X, we provide *Median Tables* and *Decile Plots*, which sort all machines in ascending order by the value of X. Think of this as generating a line, $sort(X)$ that lines up all hosts from left to right with the left-most host having the lowest value of X and the right most having the highest value of X.

Median tables. In the median tables, we divide the $sort(X)$ line at the median value for X, and we show the average number of attacks per host for the hosts to the left of the median (lower 50% of hosts) and the ones to the right of the median (upper 50% of hosts). For instance, Median Table 4.3 shows that hosts with fewer than the median number of binaries receive 2.9 attacks on average, while hosts with an above-median binary count receive 3.7 attacks on average.

Decile plots. These plots split $sort(X)$ into 10 equal-sized deciles from left to right. The first decile has the 10% of machines with the lowest values of the IV, while the tenth decile has those with the top 10%.

Kernel-Density (KD)-plots. Decile plots line up hosts by the value of an independent variable and look at the attacks counts. In contrast, kernel-density (KD) plot (e.g. Fig. 4.4) sort hosts by *how often they were attacked*: (1) hosts without attacks, (2) hosts with a few (1 or 2) attacks, and (3) hosts with a higher (≥ 3) number of attacks. For each group we plot the density function of their score of some independent variable.

Table 4.3 Average number of attacks for hosts with greater or less than the median of the number of binaries (diff. significant at $p < 0.001$)

Binary count	Gamer	Pro	SW-dev	Other	All
<Median	5.4	3.5	6.6	2.6	2.9
>Median	5.4	4.2	8.6	2.9	3.7

Statistical tests. We want to test whether the differences the upper and lower 50%, reported in the median tables, are statistically significant or whether they are due to pure chance. To check whether the means for the two groups of hosts (>*median* vs. <*median*) are statistically significant, we conduct the well-known Mann-Whitney U-test (using a Bonferroni correction to account for the number of significance tests conducted on the same data set).

In all the experiments reported in this section, the p-value is under 0.001. This is true even when the values in the median tables are identical up to the first decimal place. This high statistical significance is not surprising given our massive dataset.

4.5.2   Number of Binaries and Risk

We compare the number of attacks against machines that have less than and more than the median number of binaries via Median Table 4.3.

In the case of SW-dev hosts, there is strong evidence to support the hypothesis that the number of binaries on a host is linked to the number of attacks: developers with above-median binary counts receive 8.6 attacks on average, a 30% increase over the value for developers with fewer than the median number of binaries on their hosts. The Decile Plot in Fig. 4.3 also shows that SW-dev hosts exhibit a clear upward trend in the number of attacks as the number of binaries goes up, suggesting that a high number of binaries is associated with more attacks against hosts used for software development purposes. Moreover, we see that SW-dev machines are more heavily attacked irrespective of the decile considered, followed by gaming and pro categories.

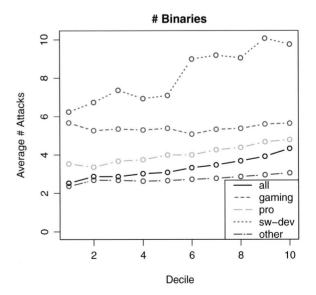

Fig. 4.3 Graph showing average number of attacks against machines in the i'th decile by number of binaries for $i = 1, \ldots, 10$

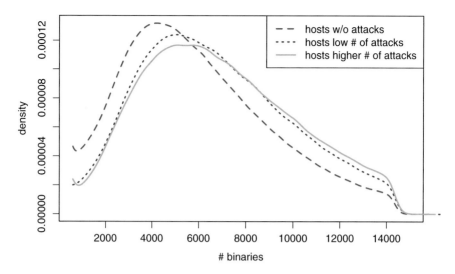

Fig. 4.4 Kernel-density plots of number of binaries

For the other categories of users, the link between the number of binaries and the risk of cyber attacks is weaker, while still statistically significant. Additionally, the slopes of the corresponding curves in Fig. 4.3 are closer to the horizontal. This suggests that there is something uniquely risky about the way software developers acquire binaries. Moreover, the more binaries they acquire, the more likely it is that they will be attacked, which suggests that the total number of binaries present on a host is an important risk factor for software developers.

Figure 4.4 shows kernel-density (KD) plots of the number of binaries per host. We see the same trend. The higher the attack count, the more the density curve shifts right, i.e. has more binaries. However, the shift is minor and there is a large overlap in the density curves. This means, that the number of binaries explains only some of the difference between the low, medium, and highly attacked machines.

4.5.3   Percentage of Low/High Prevalence and Unique Binaries and Risk

We checked if there was a difference in risk associated with low prevalence binaries (present on under 1K hosts), medium prevalence (present on 1K to 1M hosts) binaries, and high prevalence binaries (present on over 1M hosts). We also consider unique binaries, i.e. binaries we found on only one host. Median Table 4.4 shows how the number of attacks changed when we considered Low/High prevalence and unique files which were below/above the median number of files in each category.

Table 4.4 Average number of attacks for hosts with greater or less than the median of the fraction of low/high/unique binaries (diff. significant at p <0.001)

Prevalence	Gamer	Pro	SW-dev	Other	All
<Median low	2.9	2.5	4.0	2.0	2.2
> Median low	7.5	4.9	8.4	3.6	4.4
<Median high	6.1	4.6	9.0	3.5	4.2
> Median high	3.7	2.8	3.9	2.2	2.4
<Median unique	4.5	3.5	7.2	2.4	2.9
> Median unique	6.8	4.5	8.6	3.1	3.7

This table shows a clear trend—SW-dev hosts are most at risk, followed by gamer followed by pro, and this risk goes up as the percentage of low prevalence binaries increases, irrespective of which category is considered. When we look at the high prevalence binaries, the trend is reversed as we would expect. The Decile Plot of Fig. 4.5 shows the number of attacks for machines in each category by decile.

Software developers create unique executable binaries by writing and compiling programs. These executables are included in our low-prevalence binary counts but they do not pose a threat to the developers' hosts. Our analysis of the impact of unique files highlights this trend: software development hosts with above-median unique files receive only 20% more attacks, in contrast to the >100% increase when considering low-prevalence files, and the corresponding line in the decile plot is not monotonically increasing. We observe a similar trend for professional users.

The huge increase in attack numbers for gamer hosts with a high ratio of low-prevalence binaries is remarkable. From the eighth to the tenth decile, the average attack count more than doubles. The results suggest that there is a subpopulation of gamers with especially risky behavioral patterns.

These plots provide further evidence supporting the hypothesis that an increased fraction of low prevalence files on a host increases the risk of malware attack—and, as expected, as the fraction of high-prevalence files increases, the risk level decreases.

Figure 4.6 shows separate KD-plots for binaries with low, medium and high prevalence. We observe the same pattern. In the first graph, the KD-plot for low-prevalence binaries shows that every type of host (with a low, medium or high number of malware attacks) has larger numbers of attacks than in other cases.

Software Developers are at Higher Risk. Figure 4.2 shows that software developers have a higher rate of low-prevalence binaries and unique binaries, possibly because they compile binaries. They also have much higher attack rates than other groups. We wondered if the high attack rate is the result of misclassification of custom-build binaries as malware. Symantec anti-virus products detects malware using both static signatures (e.g. checking the file hash against a blacklist, scanning for strings or regular expressions in the binary content) and the behavior of a binary, observed at runtime (e.g. downloading files, modifying Windows

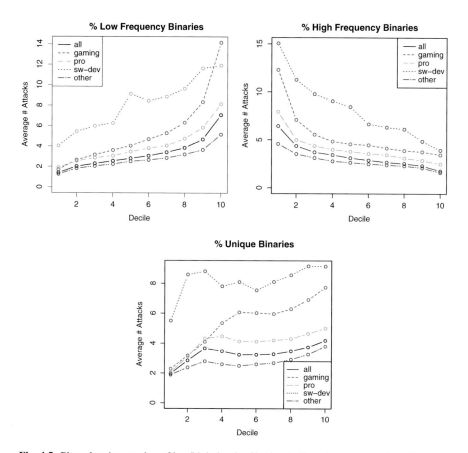

Fig. 4.5 Plots showing number of low/high density files by decile and average number of attacks by decile for deciles $1, \ldots, 10$

registry entries). Perhaps newly compiled binaries, which are unique files and look naturally suspicious to an anti-virus, further trigger behavior-detection heuristics that cause them to be reported as malicious. We therefore broke down the attack numbers separately for signature-based and behavioral virus detection. SW-dev hosts have on average 7.2 pieces of malware detected based on signatures and 0.9 pieces of malware detected by behavior. For Non-SW-Dev these numbers are 3.0 and 0.2, respectively. While behavioral detections do result in 4.5 more attack reports for software developers—as opposed to 2.4 more in the case of signature-based detections—their contribution to the overall attack averages is small. This suggests that our results are not distorted by misclassification of benign binaries as malware.

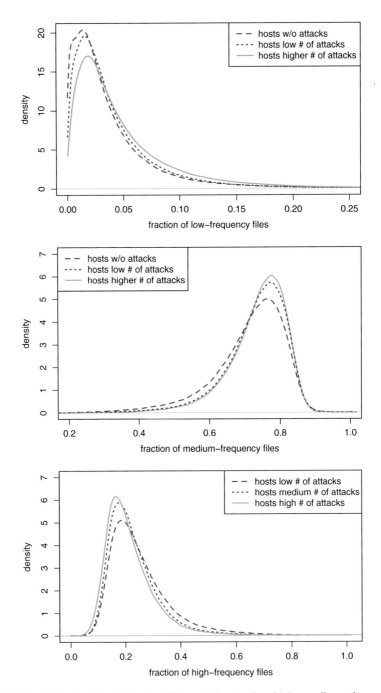

Fig. 4.6 Kernel-density plots of fraction of files with low/medium/high overall prevalence

Table 4.5 Average number of attacks for hosts with greater or less than the median of the fraction of unsigned/unsigned and unique binaries (diff. significant at $p < 0.001$)

Binaries	Gamer	Pro	SW-dev	Other	All
<Median unsigned	3.9	2.9	5.3	2.4	2.6
>Median unsigned	6.3	4.6	8.9	3.2	4.0
<Median unsigned and unique	3.5	3.5	7.0	2.4	2.9
>Median unsigned and unique	6.7	4.5	8.7	3.1	3.7

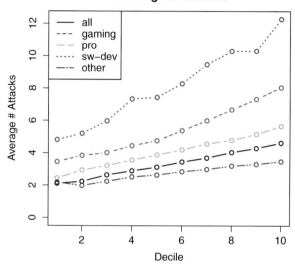

Fig. 4.7 Plot showing number of attacks for host machines in the i'th decile by number of unsigned binaries for $i = 1, \dots, 10$

4.5.4   Percentage of Unsigned Binaries and Risk

The Median Table 4.5 shows the link between the number of unsigned binaries on a host machine and the total risk. The table suggests that hosts with a larger than median percentage of unsigned binaries are more at risk than those with a less than median percentage of unsigned binaries. It also suggests that gamers are more vulnerable than Pros who in turn are less vulnerable than SW-dev. However, the higher risk associated with this feature seems comparable across the three user categories: we observe 59–68% more attacks against hosts with above-median numbers of unsigned binaries. The Decile Plot in Fig. 4.7 further substantiates this observation.

We also checked for binaries that are both unsigned and unique. As mentioned before, binaries might be unique because they are infected by a polymorphic virus that changes its own code to avoid signature-based detection. Additionally, legitimate binaries might be unique because they result from just-in-time (JIT) compilation

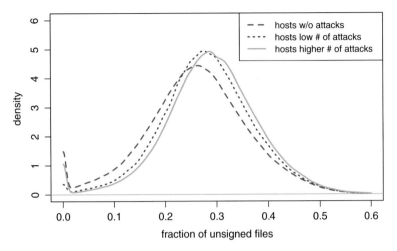

Fig. 4.8 Kernel-density plots of fraction of unsigned binaries

during installation or on the first launch or because they include customer-specific licensing information; however, such legitimate binaries are typically signed. We therefore checked whether unsigned and unique binaries are better indicators for malware attacks. For developers and professional users, this does not seem to be the case (the unique binaries that result from program compilation are not typically signed). However, for gamers the addition of the uniqueness criterion leads to a higher risk profile: 91% more attacks for hosts with above-median numbers of unique unsigned files, as opposed to 62% more attacks when considering all unsigned files.

Finally, Fig. 4.8 presents a kernel density plot showing that hosts with a higher fraction of unsigned files experience a higher occurrence of malware.

4.5.5 Percentage of Downloaded Binaries and Risk

Median Table 4.6 compares the average attack numbers of hosts on which we found more or less than the median fraction of downloaded binaries. We see a clear link between the source of binaries and attack numbers for all studied groups of the user population. This feature seems to affect our three user groups equally. The by-decile data of Fig. 4.9 shows an especially sharp increase in attack numbers for the ninth and tenth decile.

The KD plot in Fig. 4.10 shows a much better separation of the density curves for the three classes of hosts than the plots for most other IVs, suggesting that the hosts that are not attacked exhibit a different download behavior than the other hosts. In particular, for hosts with no attacks, the most frequent value in the download count distribution is 0. This clearly shows that users who refrain from downloading any executable files from the Internet tend to be safe from cyber-attacks.

Table 4.6 Average number of attacks for hosts with greater or less than the median of the fraction of downloaded binaries (diff. significant at p<0.001)

Downloaded binaries	Gamer	Pro	SW-dev	Other	All
<Median	3.7	2.8	5.5	2.2	2.4
>Median	6.4	5.0	9.8	3.4	4.2

Fig. 4.9 Plot showing number of attacks for host machines in the i'th decile by fraction of downloaded binaries for i=1,...,10

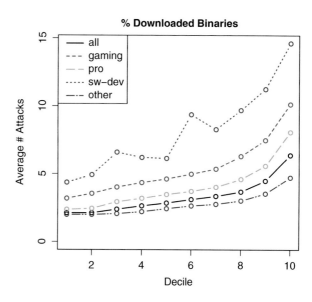

Fig. 4.10 Number of downloaded files density

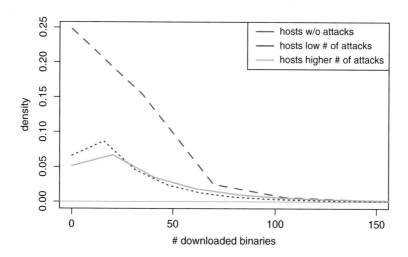

4.5.6   User Travel History and Risk

We studied the hypothesis that increased travel by a host machine increases the risk and number of attacks of that machine. Measuring travel is hard—we used the number of Internet Service Providers (ISPs) that a host has connected through as a proxy for the amount of travel by the user of that machine. Table 4.7 shows the number of attacks for machines above/below the median number of ISPs that a machine connected to. As before, SW-dev hosts receive more attacks than gamer hosts, which receive more attacks than pro hosts.

Moreover, there is a clear increase in the number of attacks on hosts that are above the median in terms of the number of ISPs that they connected from. As in the previous sections, we investigated this further by looking at the number of attacks on a decile by decile basis w.r.t. the number of ISPs that hosts connect to—the result is shown in Fig. 4.11.

The trend here is not as clear as in previous cases, and the risk does not increase as much for hosts with above-median travel histories (up to 24% more attacks, for professional users). Nonetheless, there is a discernible upward trend (especially when we get to the higher deciles). In particular, for all categories of hosts, we see that when we get to the eighth decile, there is a marked increase in the number of

Table 4.7 Average number of attacks for hosts with greater or less than the median of the number of ISPs (diff. significant at $p < 0.001$)

ISPs	Gamer	Pro	SW-dev	Other	All
<Median	4.9	3.7	7.5	2.6	3.0
>Median	5.9	4.6	8.7	3.1	3.8

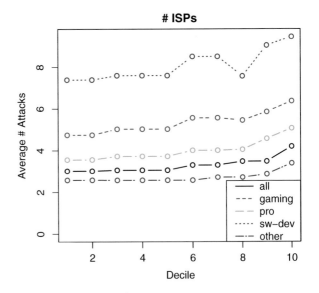

Fig. 4.11 Plot showing number of attacks for host machines in the i'th decile by number of ISPs that the machine connected to for $i = 1, \ldots, 10$

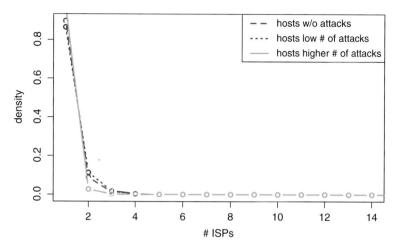

Fig. 4.12 Number of ISP density

attacks. This is further confirmed by the kernel density plot in Fig. 4.12—but the connection between number of ISPs and a higher risk of attack is weaker than in preceding sections. As the influence of the ISP count on the number of attacks is rather low and may also reflect the usage intensity of a host, we do not claim a strong connection between travel frequency and attacks.

4.6   Related Work

We considered related work in cybersecurity and data mining. However, to the best of our knowledge, we are the first to do an in-depth analysis of a complete (though cleaned) data set spanning an 8 month period.

Human Factors in Security: A growing body of research points to the importance of human behavior in creating security products. Leach [17] found that "as many as 80% of major security failures could be the result of not poor security solutions but of security behavior…".

Abraham et al. [18] specifically studied "social engineering" malware, which adopts a combination of psychological and technical ploys, with the eventual goal of luring a computer user to execute the malware. Visualization tools for security analysts is also an active research area [19].

Carlinet et al. [20] analyzed the network traffic of ADSL users to identify risk factors. They identified the usage of web and streaming increases the infection risk while for peer-to-peer and chat applications usage no such link could be established. L'evesque et al. [21] recently conducted a field study where they installed monitoring software on the computers of 50 subjects to identify risk factors for malware attacks. They primarily analyzed the types of websites their subjects visited (e.g. mp3/streaming, sport, gambling, illegal). Our analysis overlaps with theirs in two respects: number of applications/binaries and computer expertise. Like them we identified a significant relationship between attacks and application/binaries. However, in our *large-scale analysis* we also see that although the relationship is statistically significant the influence is low. Interestingly, using completely different methods, with respect to computer expertise, our results for SW-developers validate the observation of [21] that higher computer expertise is a risk factor.

Data Mining for Security: Much research has tried to model malware propagation. Staniford et al. [22, 23] analyzed the Code Red worm traces and proposed an analytical model for its propagation. They also argue that optimizations like hit-list scanning and permutation scanning can allow a worm to saturate 95% of vulnerable hosts on the Internet in less than 2 s. Papalexakis et al. [12] propose the SharkFin and GeoSplit models of spatio-temporal propagation of malware based on an analysis of the WINE data. Their system models only the total volume of malware attacks as a whole over time without considering the human behavioral aspect. In contrast, in this chapter we model the *magnitude* of malware attacks *per machine* in the context of the *machine usage* by humans. As such, our work can be also thought of as providing a fine-grained picture of human behavior characteristics that seem to be related to increased vulnerability to malware.

4.7 Discussion and Conclusion

Though humans are believed to be one of the weaker links in cybersecurity [4], hardly any work to date has focused on the relationship between the behavior of human users of machines and malware attacks on those machines. In this chapter, we report the results of an extensive analysis of Symantec's WINE dataset in which we studied 13.7B malware reports on 1.6M machines over an 8-month period. We identified a set of machine features (number of binaries, fraction of unsigned, downloaded, low prevalence, and unique binaries, and the number of ISPs the user connected from) related to behaviors of the user. For instance, these features are proxies for the tendency of users to download lots of binaries, to travel a lot, to download rare pieces of code, and to work at odd hours of the day. We grouped all users into five categories: gamers, pros, SW-dev, other and all.

Our results show that all of these variables are related to the number of pieces of malware found on their host machines at a statistically significant level ($p < 0.001$). Of these statistically significant results, the ones that we deem the most solid are ones showing that the number of malware attacks on a machine are related to the number of downloaded, unsigned, and low prevalence binaries for all categories of users. Moreover, the number of binaries on hosts are linked to number of malware attacks on hosts in the SW-developer category also at the statistically significant $p < 0.001$ level even when we account for the fact that software developers may generate binaries by compiling code they are developing.

References

1. Anderson RJ (1993) Why cryptosystems fail. In: Denning DE, Pyle R, Ganesan R, Sandhu RS, Ashby V (eds) ACM Conference on Computer and Communications Security, pp 215–227
2. Clark S, Goodspeed T, Metzger P, Wasserman Z, Xu K, Blaze M (2011) Why (special agent) Johnny (still) can't encrypt: a security analysis of the APCO project 25 two-way radio system. In: Proceedings of the 20th USENIX conference on Security, pp 4–4, USENIX Assoc
3. Whitten A, Tygar JD (1999) Why Johnny can't encrypt: A usability evaluation of PGP 5.0. In: Proceedings of the 8th USENIX Security Symposium, vol 99, McGraw-Hill, New York
4. Schneier B. Semantic attacks: The third wave of network attacks. https://www.schneier.com/crypto-gram-0010.html#1
5. Grier C, Ballard L, Caballero J, Chachra N, Dietrich CJ, Levchenko K, Mavrommatis P, McCoy D, Nappa A, Pitsillidis A, Provos N, Rafique MZ, Rajab MA, Rossow C, Thomas K, Paxson V, Savage S, Voelker GM (2012) Manufacturing compromise: the emergence of exploit-as-a-service. In: Yu T, Danezis G, Gligor VD (eds) ACM Conference on Computer and Communications Security, pp 821–832
6. Dumitras T, Shou D (2011) Toward a standard benchmark for computer security research: The Worldwide Intelligence Network Environment (WINE). In: EuroSys BADGERS Workshop, Salzburg, Austria, Apr 2011
7. Symantec Corporation (2012) Symantec Internet security threat report, vol 17. http://www.symantec.com/content/en/us/enterprise/other_resources/b-istr_main_report_2011_21239364.en-us.pdf, April 2012
8. Chau DH, Nachenberg C, Wilhelm J, Wright A, Faloutsos C (2010) Polonium: Tera-scale graph mining for malware detection. In: Proceedings of the second workshop on Large-scale Data Mining: Theory and Applications (LDMTA 2010), Washington, DC, vol 25
9. Rajab MA, Ballard L, Lutz N, Mavrommatis P, Provos N (2013) CAMP: Content-agnostic malware protection. In: Network and Distributed System Security (NDSS) Symposium, San Diego, CA
10. Cowan C (2013) Windows 8 security: Supporting user confidence. USENIX Security Symposium, invited talk, Aug 2013
11. Manadhata PK, Wing JM (2011) An attack surface metric. IEEE Trans. Software Eng., 37(3):371–386
12. Papalexakis EE, Dumitras T, Chau DH, Prakash BA, Faloutsos C (2013) Spatio-temporal mining of software adoption & penetration. In: 2013 IEEE/ACM International Conference on Advances in Social Networks Analysis and Mining
13. Bono S, Caselden D, Landau G, Miller C (2009) Reducing the attack surface in massively multiplayer online role-playing games. Security & Privacy, IEEE, 7(3):13–19
14. A.V.R. Group (2013) AVG insight: 90% of game hacks infected with malware. http://blogs.avg.com/news-threats/avg-insight-90-game-hacks-infected-malware/, Apr 2013
15. Mandiant (2013). APT1: Exposing one of china's cyber espionage units. Mandiant Whitepaper, Feb 2013
16. O'Gorman G, McDonald G (2012) The Elderwood project. Symantec Whitepaper, Oct 2012
17. Leach J (2003) Improving user security behaviour. Computers & Security, 22(8):685–692
18. Abraham S, Chengalur-Smith I (2010) An overview of social engineering malware: Trends, tactics, and implications. Technology in Society, 32(3):183–196
19. Nataraj L, Karthikeyan S, Jacob G, Manjunath BS (2011) Malware images: visualization and automatic classification. In: ACM Proceedings of the 8th International Symposium on Visualization for Cyber Security, VizSec '11, pp 4:1–4:7
20. Carlinet L, Me L, Debar H, Gourhant Y (2008) Analysis of computer infection risk factors based on customer network usage. In Emerging Security Information, Systems and Technologies, 2008. SECURWARE Aug 2008. Second International Conference on, pp 317–325

21. Lalonde L'evesque F, Nsiempba J, Fernandez JM, Chiasson S, Somayaji A (2013) A clinical study of risk factors related to malware infections. In: *Proceedings of the 2013 ACM SIGSAC Conference on Computer Communications Security*, CCS '13, New York, NY, USA, pp 97–108
22. Staniford S, Moore D, Paxson V, Weaver N (2004) The top speed of flash worms. In: Proceedings of the 2004 ACM workshop on Rapid malcode, WORM '04, New York, NY, USA, pp 33–42
23. Staniford S, Paxson V, Weaver N (2002) How to own the internet in your spare time. In: Proceedings of the 11th USENIX Security Symposium, Berkeley, CA, USA, pp 149–167, USENIX Assoc

Chapter 5
Country by Country Analysis

In this chapter, we discuss the cyber-vulnerability status of each of the 44 countries that are included in our study.

We start with a discussion of each country's national cybersecurity strategy (where available). Some countries (such as Argentina) do not appear to have released a widely available national cybersecurity strategy, though they do have cybersecurity capabilities within their country's defense forces [1, p. 5]. Other countries, including many not included in our study, have detailed national cybersecurity strategies. For instance, the European Union's Agency for Network and Information Security (ENISA) hosts a comprehensive catalog [2] of national cybersecurity strategies of many countries including ones such as Estonia, Latvia, Kenya, Namibia, Rwanda, and Uganda that are not covered in this study. ENISA has also released a document detailing how nations should develop national cybersecurity strategies [3]. This includes steps such as assessing national risk, developing clear governance structures, engaging with stakeholders, developing cybersecurity contingency plans, creating regular cybersecurity exercises, developing organizations for incident reporting, establishing public private partnerships, and more. In a similar vein, the International Telecommunications Union (ITU) has developed a detailed plan for creating national cybersecurity policies [4]. Their plan includes the creation of a legislative framework for cybersecurity, taking technical measures (including the development of risk assessment models), developing the appropriate organizations and organization reporting frameworks, building capacity in both the technical area of cybersecurity as well as in legislative and enforcement arenas, and setting up a framework for international cooperation. This last "pillar" in the ITU document recognizes the cross-border nature of cybersecurity.

© Springer International Publishing Switzerland 2015
V.S. Subrahmanian et al., *The Global Cyber-Vulnerability Report*, Terrorism, Security, and Computation, DOI 10.1007/978-3-319-25760-0_5

One common aspect of all of these cybersecurity strategies is the lack of objective cybersecurity data used in the formulation of policy. If such data, similar to that reported in this chapter, was used—it was not reported in any of documents [1–4]. This is a rather frightening thought—are national cybersecurity strategies being formulated in the absence of objective data by reputed institutions such as the United Nations Institute for Disarmament Research (UNIDIR) [1], the European Network and Information Security Agency [2, 3] and the International Telecommunication Union [3]? We do not suggest that the answer to this question is "yes"—but we do worry that in the absence of reporting of the data used in making such suggestions, these reports fall short of what policy makers need to make the best cybersecurity decisions for their country. In this chapter, we attempt to fill this void by providing objective data—at least for the years 2010 and 2011—that tells individual countries the nature of the attack vector directed at them and where they need to target their resources.

5.1 Argentina

We were unable to find a National Cybersecurity Strategy for Argentina. The closest we could come in our investigations was a document from the Organization of American States (OAS) [4] that looks at security in the Americas.

Our analysis of Argentina's cyber-vulnerability shows that the country is doing well in the sense that the percentage of attacked hosts and the average number of attacked hosts are both below the numbers one would expect for a country with Argentina's GDP. The table below shows these numbers.

Argentina	Avg. number of attacks per host	Percentage of attacked hosts
2010	7.64	0.71
2011	6.88	0.66

Compared to other countries, Argentina's percentage of attacked hosts is very large (71% in 2010 and 66% in 2011), though by no means the worst in our study. This suggests that Argentina needs to improve cyber-hygiene education and perhaps find ways to reduce the costs of deploying anti-virus software.

The most common malware that is detected in Argentina in Symantec's WINE data set is Trojans (by far the most common category of malware) with worms following in second place. Argentina is targeted slightly more heavily by misleading software and spyware than is consistent with their GDP, suggesting that education efforts may also highlight how these types of malware work (Figs. 5.1, 5.2, 5.3, 5.4, 5.5, 5.6, 5.7, and 5.8).

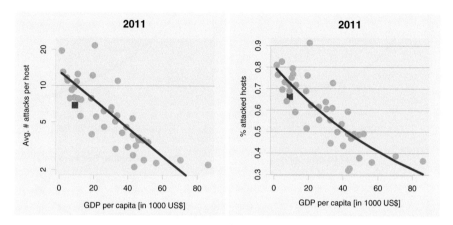

Fig. 5.1 Average number of attacks per host (*left*) and percentage of attacked host (*right*). *Blue line*: predicted values based on GDP-only model

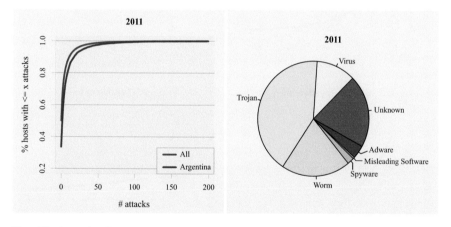

Figs. 5.2 Argentina: Empirical cumulative distribution of % of hosts with less than or equal to x attacks, and **5.3** Argentina: Distribution of attacks by type of malware

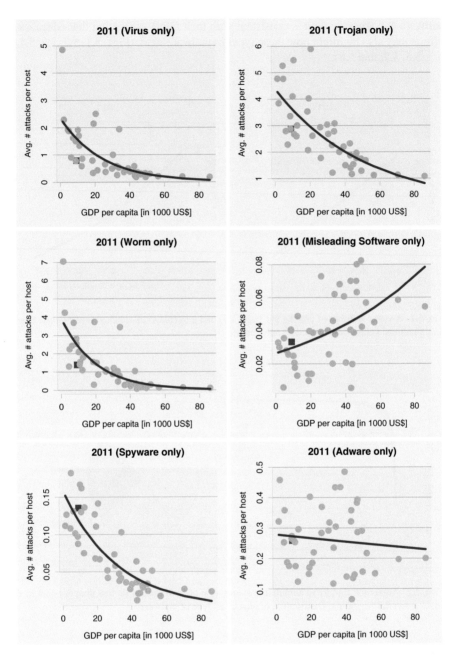

Fig. 5.4 Relationship between the GDP per capita and the average number of attacks on hosts of a country separately for virus, Trojan, worm, misleading software, spyware and adware attacks. Selected countries highlighted

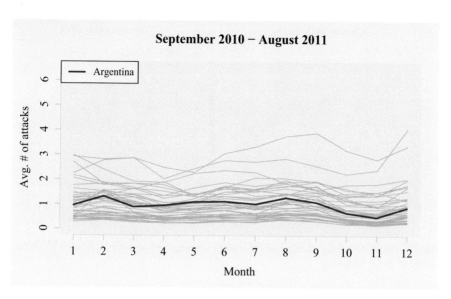

Fig. 5.5 Timeline of average monthly attacks per host by country for the 2011 time period

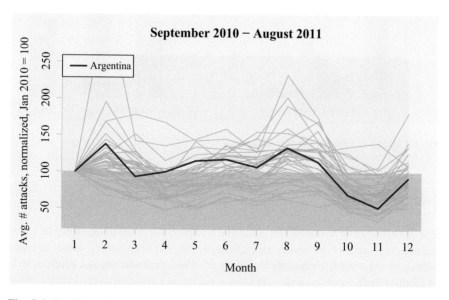

Fig. 5.6 Timeline of average monthly attacks per host by country for the 2011 time period. Normalized by attack count in September 2010

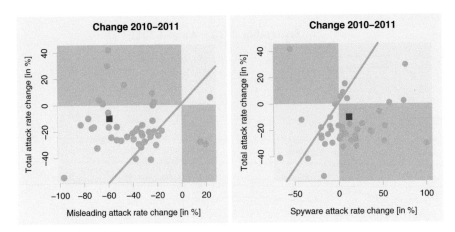

Fig. 5.7 Attack frequency changes of misleading software and spyware in relation to overall change of attack frequency

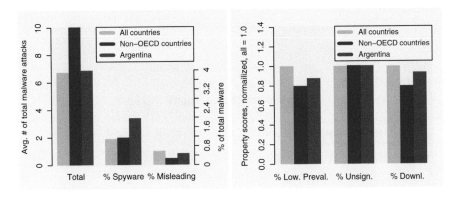

Fig. 5.8 Attack frequencies and host properties

5.2 Australia

The government of Australia has enacted a comprehensive set of measures to combat not only cyber-crime and cyber-espionage, but also to ensure the safety of individuals at risk—such as children in danger of sexual exploitation, and adults at risk of identity theft, financial fraud, and more. Australia's national cybersecurity strategy looks both at the civilian and defense perspective [5].

Institutions and Systems: [6] describes a number of organizations, systems, and measures taken by the Australian government to promote the well-being of its citizens and its economic and security interests.

- CERT Australia. CERT Australia, like many national CERTs, has the responsibility to continuously track threats worldwide, issue alerts to the Australian government, private companies, and citizens, and coordinate operations with other nations' CERTs.
- Cyber Security Operations Center (CSOC). In contrast to CERT Australia, the CSOC is an entity within Australia's defense establishment and the Australian Signals Directorate, the equivalent of the US National Security Agency. As such, CSOC is focused more on espionage and threats to Australian critical infrastructure and national security interests as a whole, rather than threats to civilian enterprises. CSOC appears to be closely linked to the Defense Signals Directorate's OnSecure effort which looks at signals intelligence—better collection and analysis of such intelligence including cyber-intelligence.
- Trusted Information Sharing Network for Critical Infrastructure Protection (TISN): TISN focuses on cybersecurity events that affect control systems operating many of Australia's critical infrastructure. It includes specific focused efforts on protecting SCADA systems that run many national utilities and other core infrastructure.
- Australian Internet Security Initiative (AISI). This initiative focuses on methods to discover bots and computers that are parts of botnets—networks of computers that are controlled by third parties who have compromised these machines to serve their ends. AISI partners with numerous Internet Service Providers and generates a large number of security alerts every single day.

International Cooperation. The Australian government recognizes the international nature of the cybersecurity problem. It is cooperating with efforts involving the United Nations, the International Telecommunications Union (ITU), the Forum for Incident Response and Security Teams, and the International Watch and Warning Network.

Education and Awareness. The Australian government made it their very first priority to ensure that all Australians are aware of cybersecurity risks to them, and are capable of taking the steps needed to ensure that their identity, privacy, and financial information is safe online. For this, the Australian Government has taken a number of steps to ensure that their citizens are well educated. These include a number of tutorials and games produced by ACMA (Australian Communications and Media Authority) to enhance awareness of cybersecurity issues. Unfortunately, the YouTube channel run by ACMA has not generated large numbers of views, so perhaps a traditional media campaign promoting these valuable online resources is needed.

Analysis. Though [5] asserts that Australia has the fifth highest incidence of malware on Australian hosts (based on a study conducted by a security vendor, Trend Micro [6]), this number does not seem to be normalized by the number of hosts present in Australia. Our results show that Australia is one of the safest countries in the world in terms of both number of attacks per host (on average) and percentage of attacked hosts in the country. In both cases, Australia ends up

being one of the top 15 most cyber secure nations in the world. The table below
shows the percentage of infected hosts in Australia and the average number of
attacks per host in Australia.

Australia	Avg number of attacks per host	Percentage of attacked hosts
2010	5.69	0.58
2011	3.38	0.49

In terms of GDP, Australia's cybersecurity performance is consistent with what
one would expect from a country with its GDP. Australian hosts are primarily hit by
Trojans, followed by worms and viruses (Figs. 5.9, 5.10, 5.11, 5.12, 5.13, 5.14,
5.15, and 5.16).

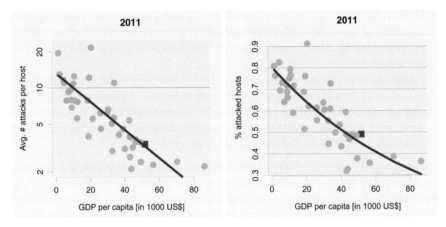

Fig. 5.9 Average number of attacks per host (*left*) and percentage of attacked host (*right*). *Blue
line*: predicted values based on GDP-only model

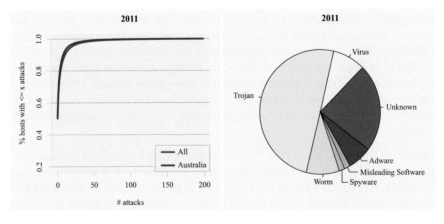

Figs. 5.10 Australia: Empirical cumulative distribution of % of hosts with less than or equal to x
attacks, and **5.11** Australia: Distribution of attacks by type of malware

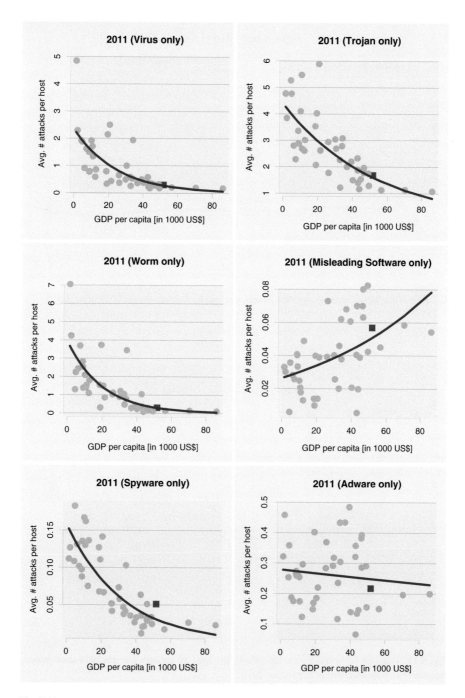

Fig. 5.12 Relationship between the GDP per capita and the average number of attacks on hosts of a country separately for virus, Trojan, worm, misleading software, spyware and adware attacks. Selected countries highlighted

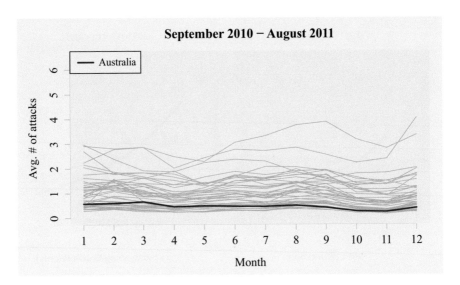

Fig. 5.13 Timeline of average monthly attacks per host by country for the 2011 time period

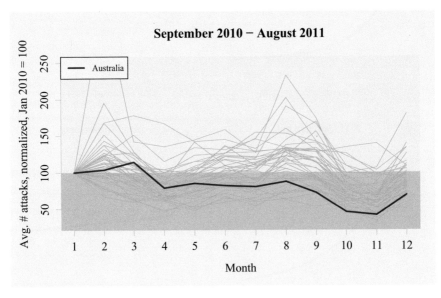

Fig. 5.14 Timeline of average monthly attacks per host by country for the 2011 time period. Normalized by attack count in September 2010

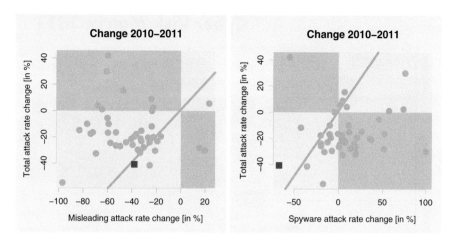

Fig. 5.15 Attack frequency changes of misleading software and spyware in relation to overall change of attack frequency

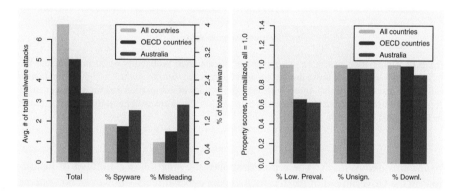

Fig. 5.16 Attack frequencies and host properties

5.3   Austria

Austria's National Cyber-Security Policy [54] was released in 2013. It describes the creation of both a set of institutions as well as a set of processes and steps to enhance Austria's cybersecurity. Specifically, the policy identifies a series of threats, along with an assessment of the probability that the threat will occur and the consequences (Fig. 5.17).

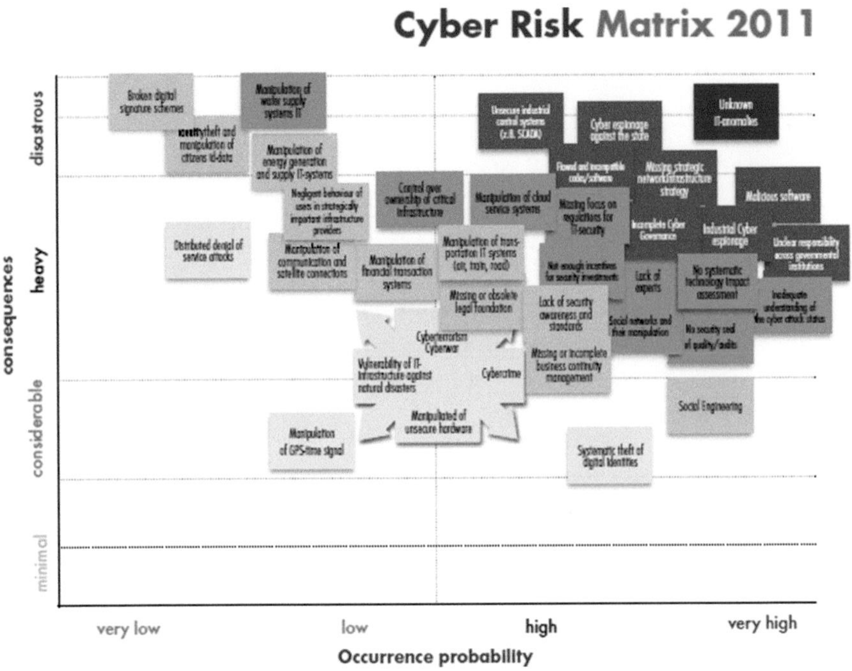

Fig. 5.17 Cyber-Risk Matrix showing types of threats, their probability of occurrence, and the (adverse) consequences if those threats are not counteracted. Entries in the top right of the matrix show the worst case threats—those with high probability of occurrence and high consequences (Reproduced from: Austrian Cyber Security Strategy [54], Bundeskanzleramt, Bundesministerium fur Inneres, 2013, Annex A, page 18)

Institutions and Systems: The report describes the creation of a number of institutions in the Government of Austria focusing on cyber-threats—however, no systems are specifically mentioned. These institutions include:

• Cyber Security Strategy Group (CSSG) based in the Federal Chancellery that includes cybersecurity officials from Austria's National Security Council, Austria's Chief Information Officer, as well as various liaison officers. The CSSG is charged with generating periodic cybersecurity "pictures", reflecting both incident reports and related information. In addition, it is charged with providing a mix of carrots and sticks to deter cyber-adversaries who are non-state actors. The CSSG also formulates a Cyber Communication Strategy focused on raising awareness and improving education related to cyber-hygiene within Austria.

- Operational Coordination Structure (OCS) is based in the Ministry of the Interior and is responsible for running and coordinating between a number of Computer Emergency Response Teams (CERTs) including ones associated with the government, military, and the private sector. This also includes a Cyber Security Competence Center which is engaged in cybersecurity related policing activities.
- Cyber Crisis Management teams include operators from various infrastructure-related authorities who have the goal of protecting national infrastructure.
- Austrian Cyber Security Platform: This is a public private partnership that runs programs linking private entities, academia, and the government, so as to share information about cyber-threats. This entity provides input to the CSSG.

International Cooperation. The Austrian government recognizes the international nature of the cybersecurity problem. It commits to participating actively in the EU Cyber Security Strategy, the Convention on Cybercrime of the Council of Europe, the NATO Partnership for Peace. In addition, the Austrian government proposes to conclude appropriate bilateral agreements as needed and to run periodic multi party cybersecurity training and response exercises.

Education and Awareness. Austria has also made a strong effort to build out a strong cybersecurity awareness program. In this program, they have developed a strong effort to include cyber-hygiene best practices into school curricula, improve cyber-security education at universities, help train government officials in cybersecurity, and help train IT staff responsible for critical infrastructure to recognize and report cybersecurity incidents.

Analysis. Our assessment of Austria, shown in detail below, shows that Austria is performing exactly at the level we would expect it to perform based on factors such as its GDP. In fact, Austria is one of the safest countries in the world from a cyber-vulnerability perspective as is illustrated in the following table.

Austria	Avg number of attacks per host	Percentage of attacked hosts
2010	4.17	0.53
2011	3.17	0.47

Most malware attacking hosts in Austria are Trojans, followed by viruses and worms. This suggests that greater emphasis on education that helps users protect against these types of malware is advisable. However, we noted that Austrians' behaviors involve downloading low prevalence binaries and unsigned binaries at a higher rate than other OECD countries. This leads to increased vulnerability—and perhaps education efforts in Austria can better address the risks of downloading such binaries (Figs. 5.18, 5.19, 5.20, 5.21, 5.22, 5.23, 5.24, and 5.25).

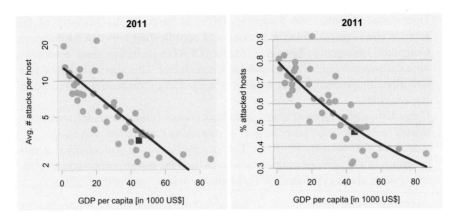

Fig. 5.18 Average number of attacks per host (*left*) and percentage of attacked host (*right*). *Blue line*: predicted values based on GDP-only model

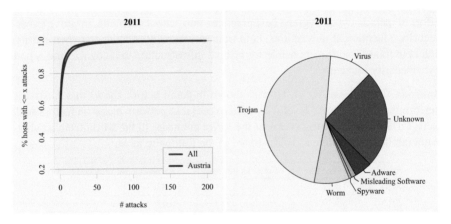

Figs. 5.19 Austria: Empirical cumulative distribution of % of hosts with less than or equal to x attacks, and **5.20** Austria: Distribution of attacks by type of malware

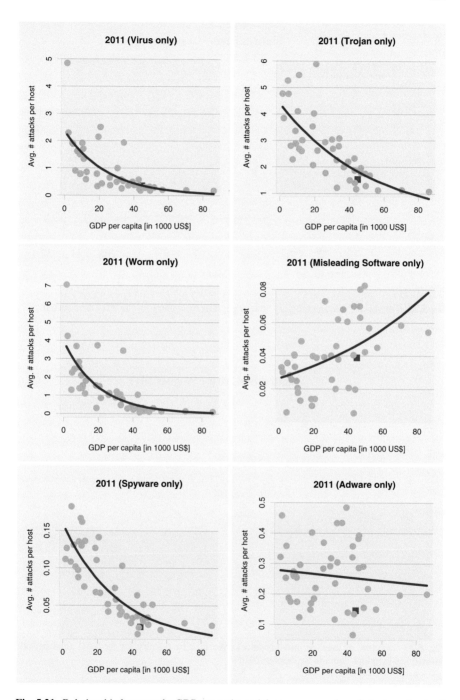

Fig. 5.21 Relationship between the GDP per capita and the average number of attacks on hosts of a country separately for virus, Trojan, worm, misleading software, spyware and adware attacks. Selected countries highlighted

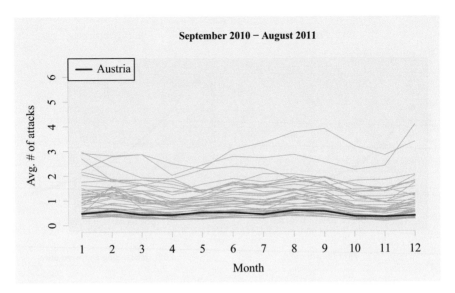

Fig. 5.22 Timeline of average monthly attacks per host by country for the 2011 time period

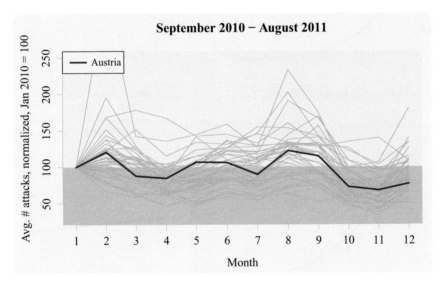

Fig. 5.23 Timeline of average monthly attacks per host by country for the 2011 time period. Normalized by attack count in September 2010

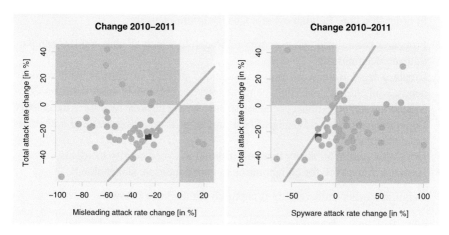

Fig. 5.24 Attack frequency changes of misleading software and spyware in relation to overall change of attack frequency

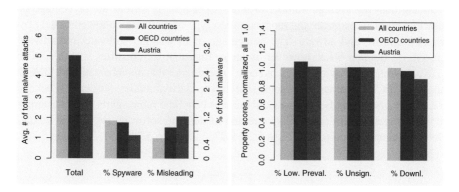

Fig. 5.25 Attack frequencies and host properties

5.4 Belgium

The Belgian National Cyber Security Strategy is documented in two documents—one [7] is entirely in French, while the other [8] is a document published by a chamber of commerce. Belgium's primary concerns focus on corporate and state-sponsored espionage. In particular, IT forms a major part of Belgium's economy—and because of this, there is deep concern that the country is vulnerable to cyber-attacks.

In order to secure Belgian cyber-space, Belgium plans to:

- Propose a centralized cybersecurity approach in which all stakeholders can participate meaningfully
- Create a legal framework that balances the privacy of citizens with security needs, while at the same time, being adaptive to previously unseen threats
- Monitoring activity within cyberspace so as to ensure physical security in the country
- Better education and advising of best practices for securing personal and enterprise machines
- Develop a central authority that will respond to cybersecurity incidents
- Ensure a framework for collaboration between multiple stakeholders.

Belgium's cyber-vulnerability is captured via the table below.

Belgium	Avg number of attacks per host	Percentage of attacked hosts
2010	4.87	0.55
2011	3.88	0.49

We see from this table that from 2010 to 2011, Belgium did an excellent job in reducing both the average number of attacks per host as well as the percentage of attacked hosts. Trojans are the most common type of malware, followed closely by viruses, worms, and adware. The percentage of misleading software and adware present on Belgian hosts, however, is larger than that for countries with a similar GDP. In terms of behavior, Belgians seem to be smarter than members of both OECD and non-OECD countries in the sense that they tend to exhibit less susceptible behavior. This suggests that Belgium should perhaps be a little more aggressive in training their population to the risks of adware and misleading software. But on the whole, Belgians seem well protected (Figs. 5.26, 5.27, 5.28, 5.29, 5.30, 5.31, 5.32, and 5.33).

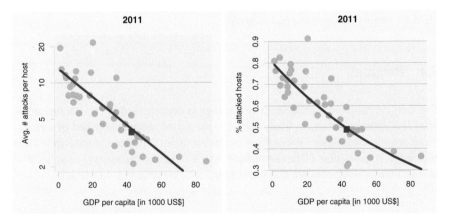

Fig. 5.26 Average number of attacks per host (*left*) and percentage of attacked host (*right*). *Blue line*: predicted values based on GDP-only model

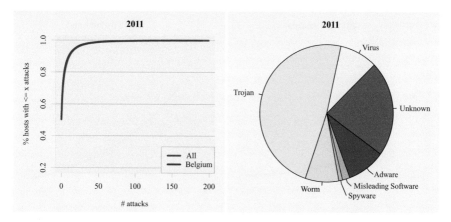

Figs. 5.27 Belgium: Empirical cumulative distribution of % of hosts with less than or equal to x attacks, and **5.28** Belgium: Distribution of attacks by type of malware

5.5   Brazil

Like many countries in Latin America, Brazil does not seem to have evolved a widely accepted national cybersecurity strategy, though the importance of cyberse-curity has certainly been recognized within their military [10]. Nonetheless, Brazil has developed the Brazilian Internet Steering Committee (CGI.br) and a Computer Emergency Response Team (CERT.br).

CGI.br is a multi-stakeholder organization run by representatives from the Brazilian Federal Government, businesses, NGOs, and universities. Brazil's CERT is a part of CGI.br's Network Information Center which also runs domain registra-tion services in Brazil and manages Internet exchange points.

Brazil's CERT is a robust and forward looking entity. Amongst other things, they run 36 Computer Security Incident Response Teams or CSIRTs [10] that are distrib-uted across the country. In order to gather better intelligence on cyber-threats to Brazil, the Brazilian CERT also runs a network of honeypots across the nation [11] and are part of the HoneyNet Project (www.honeynet.org), a global network of indi-viduals interested in collecting malware related information around the world. Brazil's CERT also maintains close connections with other CERTs around the world including ones in the US, Netherlands, and Australia. They run courses in many aspects of cybersecurity—both for cybersecurity experts and for schools and col-leges. Interestingly, they have produced cartoon-style educational videos related to anti-spam efforts (http://www.antispam.br/videos/english/) along with broader cybersecurity education efforts focused on kids (http://cartilha.cert.br/). Brazil has also carefully studied cyber-threats to critical infrastructure [55].

Brazil's cyber-vulnerability is captured via the table below.

Brazil	Avg number of attacks per host	Percentage of attacked hosts
2010	8.27	0.77
2011	7.82	0.77

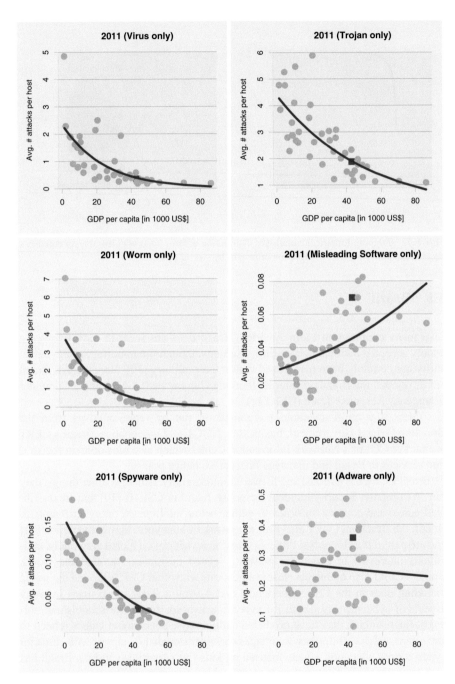

Fig. 5.29 Relationship between the GDP per capita and the average number of attacks on hosts of a country separately for virus, Trojan, worm, misleading software, spyware and adware attacks. Selected countries highlighted

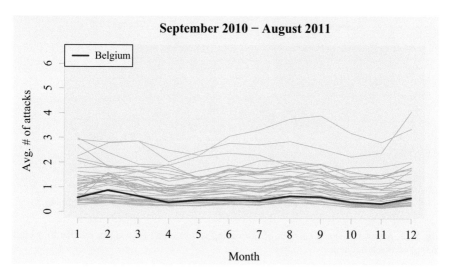

Fig. 5.30 Timeline of average monthly attacks per host by country for the 2011 time period

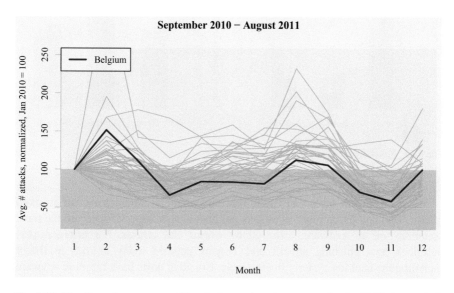

Fig. 5.31 Timeline of average monthly attacks per host by country for the 2011 time period. Normalized by attack count in September 2010

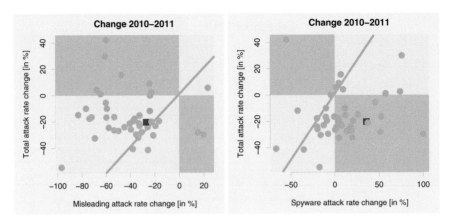

Fig. 5.32 Attack frequency changes of misleading software and spyware in relation to overall change of attack frequency

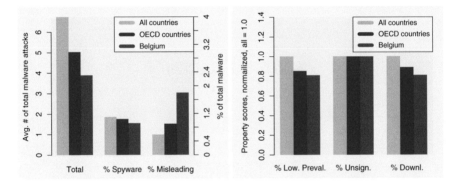

Fig. 5.33 Attack frequencies and host properties

We see from this table that Brazil has, on average, a large number of attacked hosts (77%) and the number of attacks per host is also large. Trojans are the most common type of malware, followed by worms and viruses. For instance, [11] describes how phishing attacks and Trojans targeted many entities in Brazil including the banking system. Interestingly, Brazilian hosts have far more spyware than do hosts in both OECD and non-OECD countries, suggesting that education about the risks of spyware may be worth contemplating—some steps have already been taken in this regard by the Brazilian CERT. At the same time, Brazilians have a higher tendency to download low-frequency binaries than other non-OECD countries. Moreover, the percentage of binaries on Brazilian hosts that are downloaded also significantly exceed the average for non-OECD countries. This suggests that focusing educational programs on improving web browsing and downloading behavior by Brazilians may pay appropriate dividends (Figs. 5.34, 5.35, 5.36, 5.37, 5.38, 5.39, 5.40, and 5.41).

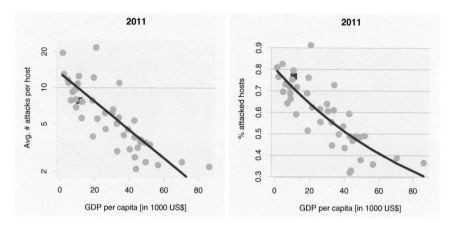

Fig. 5.34 Average number of attacks per host (*left*) and percentage of attacked host (*right*). *Blue line*: predicted values based on GDP-only model

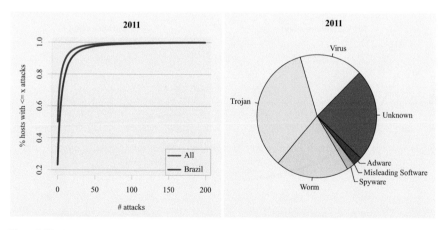

Figs. 5.35 Brazil: Empirical cumulative distribution of % of hosts with less than or equal to x attacks, and **5.36** Brazil: Distribution of attacks by type of malware

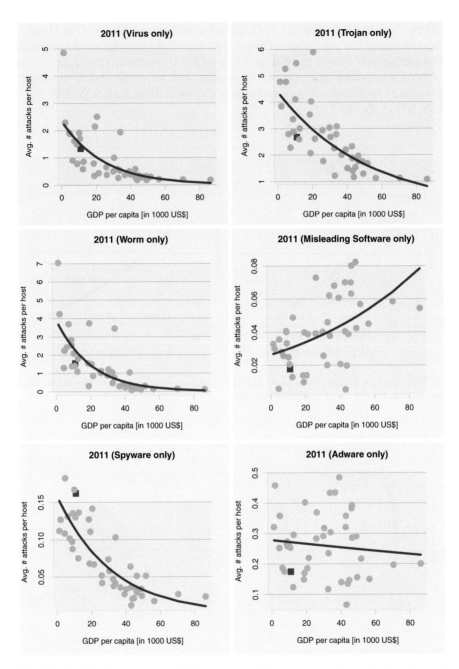

Fig. 5.37 Relationship between the GDP per capita and the average number of attacks on hosts of a country separately for virus, Trojan, worm, misleading software, spyware and adware attacks. Selected countries highlighted

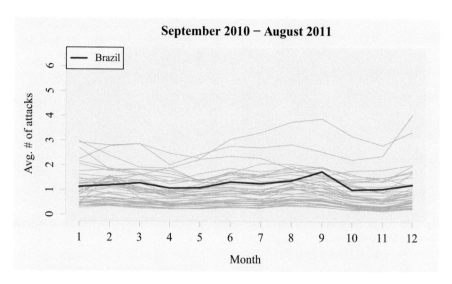

Fig. 5.38 Timeline of average monthly attacks per host by country for the 2011 time period

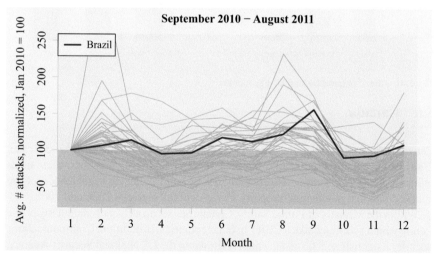

Fig. 5.39 Timeline of average monthly attacks per host by country for the 2011 time period. Normalized by attack count in September 2010

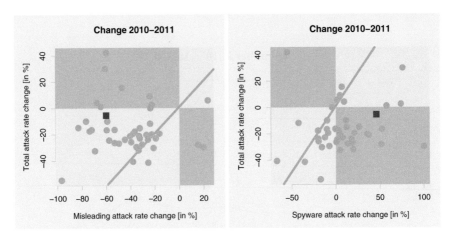

Fig. 5.40 Attack frequency changes of misleading software and spyware in relation to overall change of attack frequency

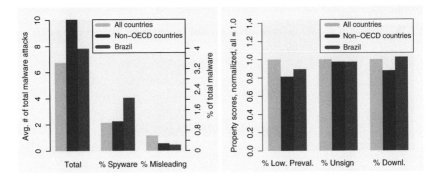

Fig. 5.41 Attack frequencies and host properties

5.6   Canada

Canada's National Cyber Security Strategy [12] recognizes (even in 2008), that 2/3rd of Canadians used online banking services and that 59% filed their taxes online in 2009. [12] further asserts that 86% of large Canadian organizations were attacked and that intellectual property losses doubled between 2006 and 2008.

Within the Ministry of Public Safety, Canada has set up a Cyber Incident Response Centre which acts much like a CERT in other countries. In addition, Communications Security Establishment Canada, which is similar to the US National Security Agency, is charged with security government networks and infrastructure. The Canadian Security Intelligence Service (similar to the CIA in the US) is charged with monitoring both home-grown and overseas cyber-threats. The Royal Canadian Mounted Police has also set up an Integrated Cyber Crime Fusion Centre.

The Canadian cybersecurity strategy recognizes that partnership is essential and advocates partnership between the national government, state government, businesses, and universities.

Canada's cyber-vulnerability is captured via the following table.

Canada	Avg number of attacks per host	Percentage of attacked hosts
2010	5.38	0.56
2011	3.66	0.48

Over half of the malware running on Canadian hosts are Trojans, with viruses and worms following in second and third place, respectively. Interesting, Canadians are far more vulnerable to misleading software (such as fake anti-virus software or fake disk cleanup utilities) than would be expected from their GDP. This suggests that better education involving fake software may be helpful in improving Canada's cybersecurity. Similarly, adware attacks on Canadian hosts also seem larger than would be expected from Canada's GDP—this too might be a target for cyber-education programs in Canada (Figs. 5.42, 5.43, 5.44, 5.45, 5.46, 5.47, 5.48, and 5.49).

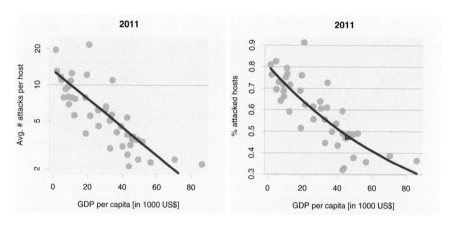

Fig. 5.42 Average number of attacks per host (*left*) and percentage of attacked host (*right*). *Blue line*: predicted values based on GDP-only model

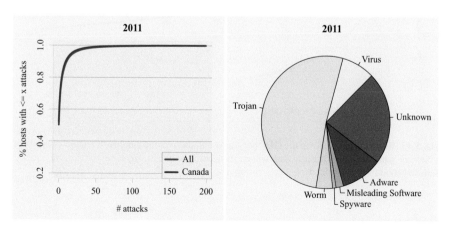

Figs. 5.43 Canada: Empirical cumulative distribution of % of hosts with less than or equal to x attacks, and **5.44** Canada: Distribution of attacks by type of malware

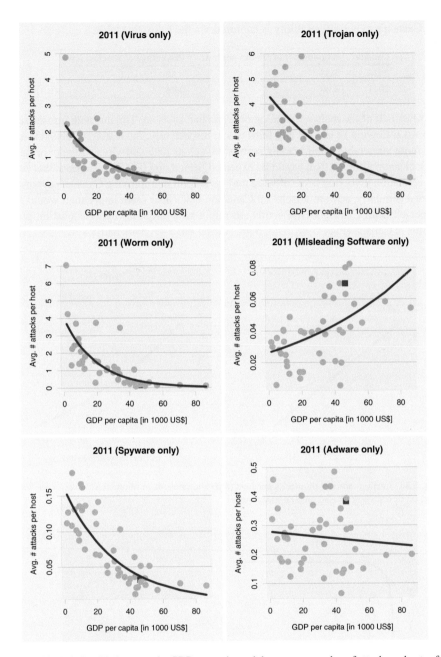

Fig. 5.45 Relationship between the GDP per capita and the average number of attacks on hosts of a country separately for virus, Trojan, worm, misleading software, spyware and adware attacks. Selected countries highlighted

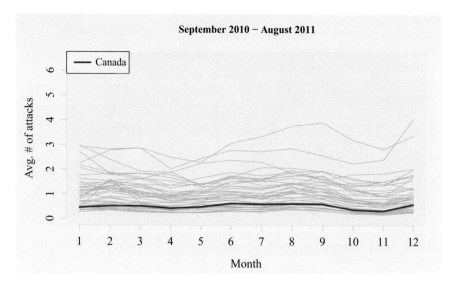

Fig. 5.46 Timeline of average monthly attacks per host by country for the 2011 time period

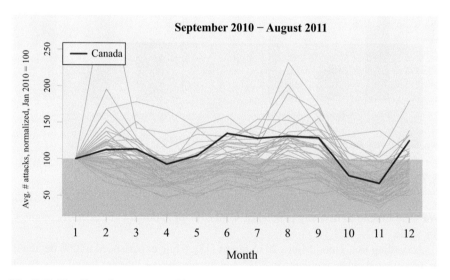

Fig. 5.47 Timeline of average monthly attacks per host by country for the 2011 time period. Normalized by attack count in September 2010

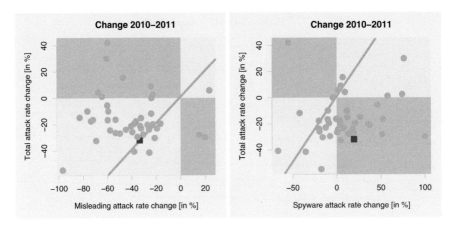

Fig. 5.48 Attack frequency changes of misleading software and spyware in relation to overall change of attack frequency

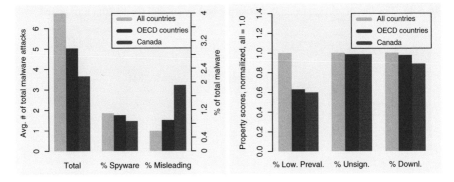

Fig. 5.49 Attack frequencies and host properties

5.7   Chile

We were unable to find any official or semi-official national cybersecurity strategy for Chile.

According to a Trend Micro 2013 report [13], Chile experienced a 22% decrease in investigated cyber-incidents in 2012 and the number of Internet-based wire fraud incidents decreased by 122% (these incidents included phishing attacks). These numbers were reported by Chile's Federal Police—however, we note that the number of "investigated" cyber-incidents could be unrelated to the total number of attacks found by us in Chilean hosts.

Our data on Chile is summarized in the table below.

Chile	Avg number of attacks per host	Percentage of attacked hosts
2010	7.23	0.70
2011	7.65	0.72

Based on Symantec's WINE data, it would appear that on whole, Chile actually became a little more cyber-vulnerable in 2011 than in 2010. During this time frame, the malware primarily found on Chilean hosts is Trojans, followed by worms and viruses. We noted that Chileans are much more vulnerable to misleading software (e.g. fake anti-virus software, fake disk cleaning utilities, etc.) than other countries with a similar GDP. This suggests that better cyber-education efforts on how to recognize misleading software is critical. Adware and spyware were also at slightly elevated levels compared to other nations with similar GDPs (Figs. 5.50, 5.51, 5.52, 5.53, 5.54, 5.55, 5.56, and 5.57).

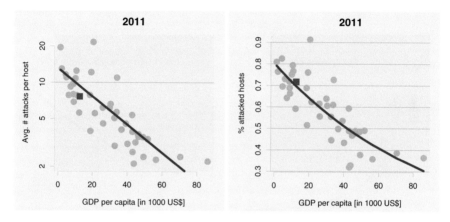

Fig. 5.50 Average number of attacks per host (*left*) and percentage of attacked host (*right*). *Blue line*: predicted values based on GDP-only model

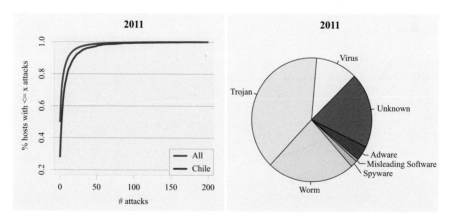

Figs. 5.51 Chile: Empirical cumulative distribution of % of hosts with less than or equal to x attacks, and **5.52** Chile: Distribution of attacks by type of malware

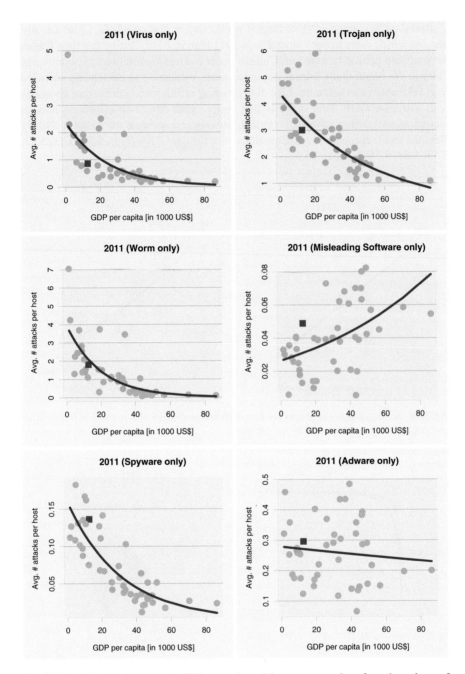

Fig. 5.53 Relationship between the GDP per capita and the average number of attacks on hosts of a country separately for virus, Trojan, worm, misleading software, spyware and adware attacks. Selected countries highlighted

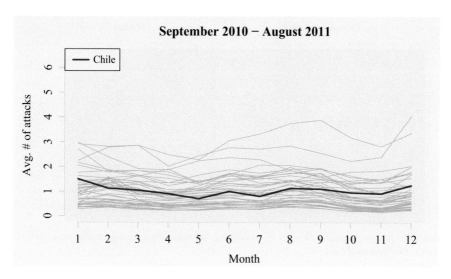

Fig. 5.54 Timeline of average monthly attacks per host by country for the 2011 time period

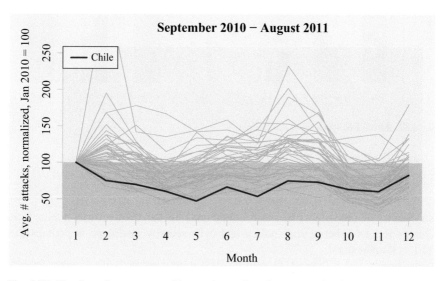

Fig. 5.55 Timeline of average monthly attacks per host by country for the 2011 time period. Normalized by attack count in September 2010

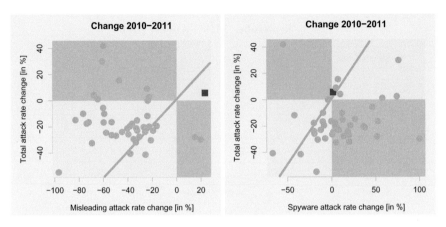

Fig. 5.56 Attack frequency changes of misleading software and spyware in relation to overall change of attack frequency

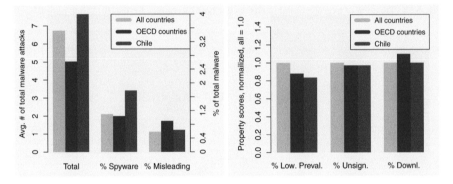

Fig. 5.57 Attack frequencies and host properties

5.8   China

No country is the subject of more controversy in the cybersecurity realm than China—yet, its cybersecurity policy and cybersecurity landscape is perhaps the least well understood.

According to Li Zhang [14] who heads a think tank in Beijing, cyber-attacks targeting China during the first half of 2012 were mostly from the USA, Japan, and South Korea. While the data reported in this book does not allow us to attribute attacks, it does allow us to definitively say that China was one of the most heavily attacked victims of cyber-attacks, irrespective of whether we measure attacks by the percentage of attacked hosts, or by the average number of malware found on hosts in the country. Zhang goes on to assert that the PLA's military networks suffered more than 80,000 attacks originating from outside China.

The only detailed external study of China's cybersecurity policy is that of Chang [15] who asserts that China's cybersecurity policy is heavily shaped by the need for the Chinese Communist Party to maintain its grip on power. She asserts that Chinese cybersecurity policy is formulated at the very highest level (President Xi Jinping's office) and that relevant Chinese institutions support this goal by using cybersecurity for economic growth (via industrial espionage), targeting sources of dissent within China, preparing for future military conflict with potential cybersecurity attacks, understanding the cybersecurity strategies of its adversaries, and come up with explanations of China's government cybersecurity policies and actions.

To this end, China has set up many different organizations shown in Fig. 5.58 below (reproduced from [15]).

Whatever China's intentions may or may not be, we are able to report some comprehensive statistics on the state of China cyberspace.

China	Avg number of attacks per host	Percentage of attacked hosts
2010	12.91	0.88
2011	11.63	0.83

Interestingly, China saw a 10% drop in the average number of attacks per host in 2011, as compared to 2010, and concurrently saw a drop of over 5% in the percentage of attacked hosts in the country. Nonetheless, both numbers are large, amongst the highest of the 44 countries in our study, suggesting that Chinese consumer hosts are still very cyber-vulnerable.

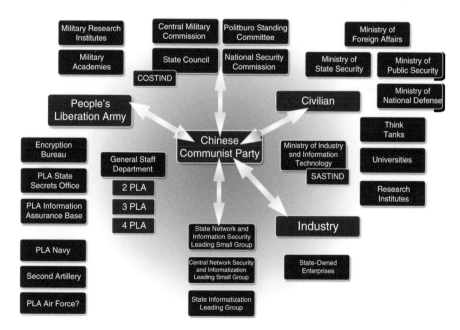

Fig. 5.58 Chart showing Stakeholders and Organizations Involved in China's National Cybersecurity Strategy (reproduced from Amy Chang, Warring State: China's Cybersecurity Strategy, Center for New American Security, p. 15, Dec. 2014.)

Of the types of malware that targets Chinese hosts, Trojans dominate, followed by viruses and worms. From a behavioral perspective, Chinese users tend to have a higher percentage of downloaded binaries and a higher percentage of low-prevalence binaries. A strategy for better cyber-health in China could well focus on increasing education about the risks of downloading strange software and in generally maintaining awareness of the risks of downloading software from the web (Figs. 5.59, 5.60, 5.61, 5.62, 5.63, 5.64, 5.65, and 5.66).

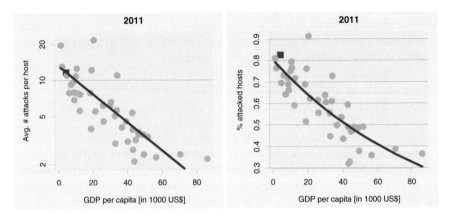

Fig. 5.59 Average number of attacks per host (*left*) and percentage of attacked host (*right*). *Blue line*: predicted values based on GDP-only model

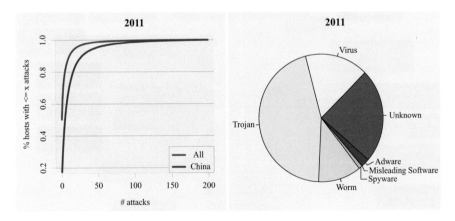

Figs. 5.60 China: Empirical cumulative distribution of % of hosts with less than or equal to x attacks, and **5.61** China: Distribution of attacks by type of malware

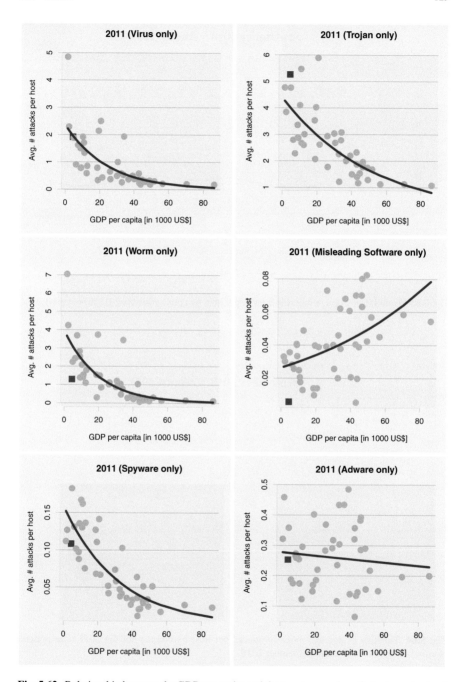

Fig. 5.62 Relationship between the GDP per capita and the average number of attacks on hosts of a country separately for virus, Trojan, worm, misleading software, spyware and adware attacks. Selected countries highlighted

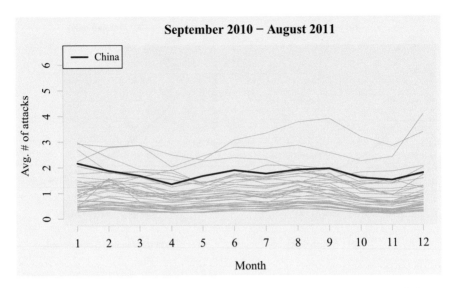

Fig. 5.63 Timeline of average monthly attacks per host by country for the 2011 time period

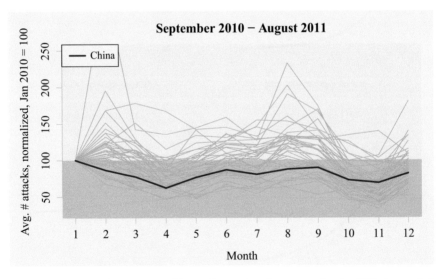

Fig. 5.64 Timeline of average monthly attacks per host by country for the 2011 time period. Normalized by attack count in September 2010

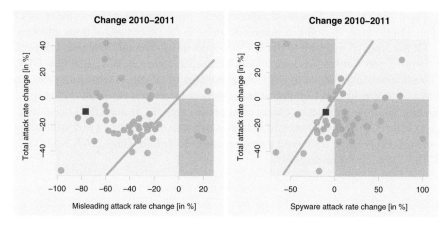

Fig. 5.65 Attack frequency changes of misleading software and spyware in relation to overall change of attack frequency

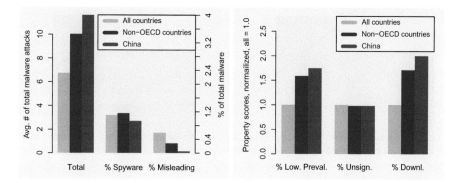

Fig. 5.66 Attack frequencies and host properties

5.9   Colombia

The principal organizations charged with implementing cybersecurity in Colombia are the National Planning Department and the Ministry for Information Technology and Communication.

According to [16], Colombia has created a number of cybersecurity organizations including a variety of CERTs and Incident Response Teams. These organizations function in a manner similar to those in other countries. The police have a cybersecurity center whose goal is to follow up on cybercrimes (e.g. crimes against children).

Colombia's cybersecurity profile is captured via the following table.

Colombia	Avg number of attacks per host	Percentage of attacked hosts
2010	9.38	0.77
2011	7.85	0.73

From 2010 to 2011, Colombia experienced a significant reduction in the number
of attacks per host, as well as a smaller decrease in the percentage of hosts attacked.

Of the types of malware found in Colombian hosts, Trojans dominate, followed
by worms and viruses. In comparison with other countries with a similar GDP,
Colombia is doing very well, with lower numbers of attacks in most categories. In
terms of user behavior, Colombian hosts show a larger percentage of downloaded
binaries and low prevalence binaries than both non-OECD and OECD countries,
suggesting that cyber-education that stresses that care must be taken in downloading
software, especially rarely used software, could be helpful (Figs. 5.67, 5.68, 5.69,
5.70, 5.71, 5.72, 5.73, and 5.74).

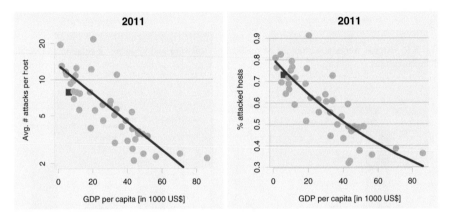

Fig. 5.67 Average number of attacks per host (*left*) and percentage of attacked host (*right*). *Blue
line*: predicted values based on GDP-only model

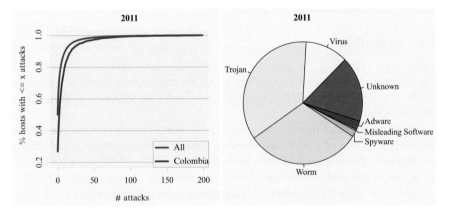

Figs. 5.68 Colombia: Empirical cumulative distribution of % of hosts with less than or equal to x
attacks, and **5.69** Colombia: Distribution of attacks by type of malware

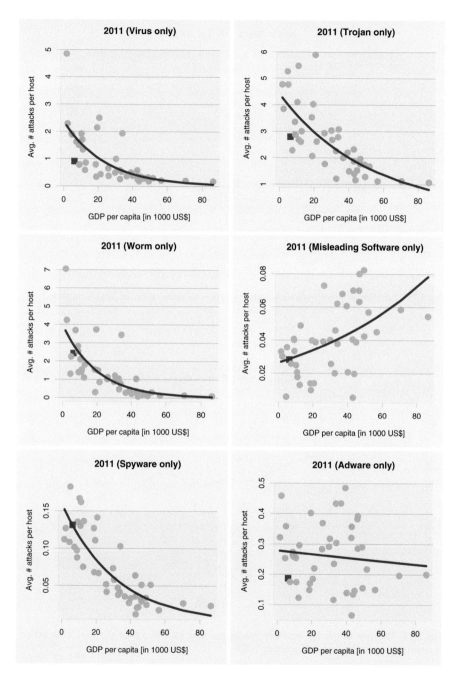

Fig. 5.70 Relationship between the GDP per capita and the average number of attacks on hosts of a country separately for virus, Trojan, worm, misleading software, spyware and adware attacks. Selected countries highlighted

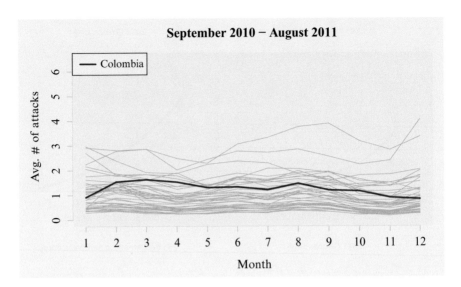

Fig. 5.71 Timeline of average monthly attacks per host by country for the 2011 time period

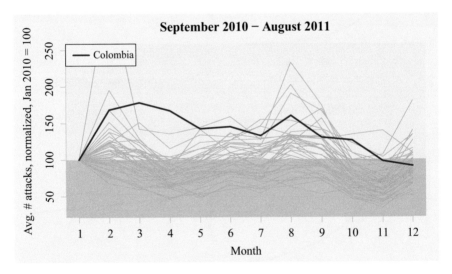

Fig. 5.72 Timeline of average monthly attacks per host by country for the 2011 time period. Normalized by attack count in September 2010

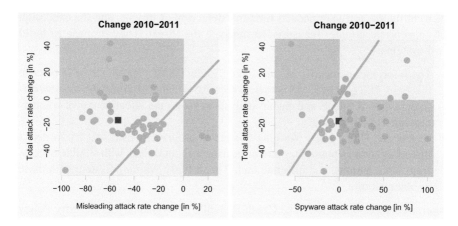

Fig. 5.73 Attack frequency changes of misleading software and spyware in relation to overall change of attack frequency

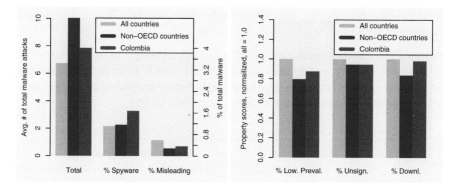

Fig. 5.74 Attack frequencies and host properties

5.10 Czech Republic

The Czech Republic's National Cyber Security Policy [9] echoes many of the same measures described in other nations' cybersecurity policies.

Institutions and Systems: The Czech Republic proposes to build only one institution, namely a Computer Emergency Response Team (CERT) whose task will be to build the software needed for early detection of threats, and the execution of appropriate counter-measures to mitigate the threat. They propose to build two systems.

- The Information Security Management System (ISMS) appears to be in place at this time. The goal of ISMS is to help security officials examine and monitor different types of threats and enable analysts to assess the effectiveness of counter-measures to the threat.
- A national Threat Early Warning System is proposed which will try to detect threats as soon as possible, share information about threats with similar systems from other nations, and integrate their Threat Early Warning System with those of other nations.

International Cooperation: The Czech Republic's National Cyber Security Policy recognizes the cross border nature of the cybersecurity threat and expresses the desire to cooperate both at the legislative and computational levels with international entities, primarily the EU and NATO.

Education and Awareness. The Czech Republic recognizes the need to have public private partnerships because of the nature of the ownership of many entities linked to critical infrastructure (e.g. banking, energy, utilities sectors). They also recognize the need for interaction with educational entities.

Analysis. The Czech Republic is batting well above its weight in terms of cybersecurity in their country. As is evident from the table below, both the number of attacks per host (on average) and the percentage of hosts that were attacked are significantly below what one would expect for a country with their GDP.

Czech Republic	Avg number of attacks per host	Percentage of attacked hosts
2010	4.62	0.57
2011	3.94	0.52

Trojans are the principal form of malware detected on machines in the Czech Republic with viruses and worms closely matched for second and third place respectively. The Czech Republic does seem to have more spyware per machine on average than other OECD countries—so educating the Czech population on how spyware gets downloaded onto machines and how to practice good cyber-hygiene with a focus on spyware may be a good mechanism to reduce this type of attack (Figs. 5.75, 5.76, 5.77, 5.78, 5.79, 5.80, 5.81, and 5.82).

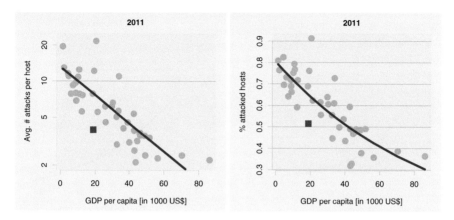

Fig. 5.75 Average number of attacks per host (*left*) and percentage of attacked host (*right*). *Blue line*: predicted values based on GDP-only model

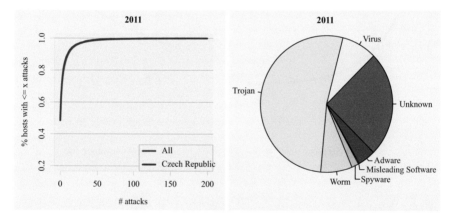

Figs. 5.76 Czech Republic: Empirical cumulative distribution of % of hosts with less than or equal to x attacks, and **5.77** Czech Republic: Distribution of attacks by type of malware

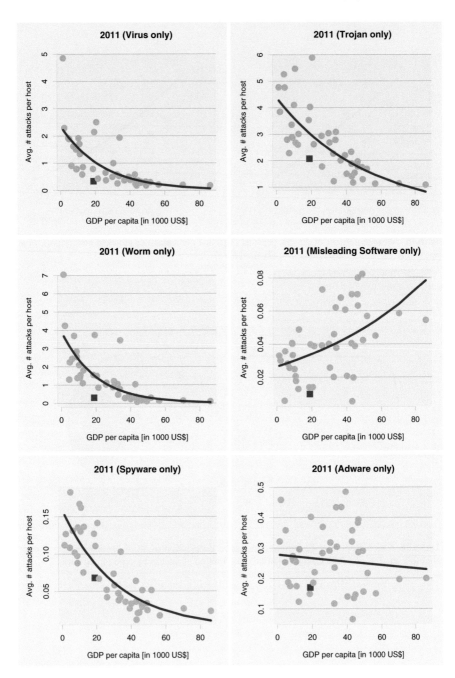

Fig. 5.78 Relationship between the GDP per capita and the average number of attacks on hosts of a country separately for virus, Trojan, worm, misleading software, spyware and adware attacks. Selected countries highlighted

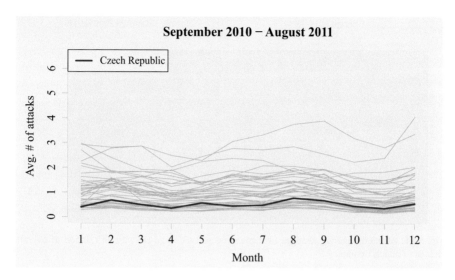

Fig. 5.79 Timeline of average monthly attacks per host by country for the 2011 time period

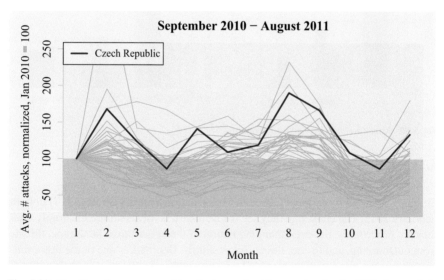

Fig. 5.80 Timeline of average monthly attacks per host by country for the 2011 time period. Normalized by attack count in September 2010

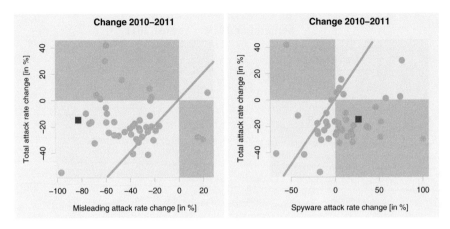

Fig. 5.81 Attack frequency changes of misleading software and spyware in relation to overall change of attack frequency

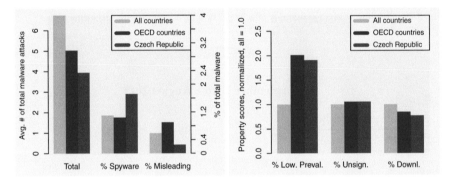

Fig. 5.82 Attack frequencies and host properties

5.11   Denmark

Denmark released its national Cyber and Information Security Strategy [17] only in February 2015. The challenge they see for Denmark involves failure to comply with security procedures, the ever changing nature of the cyber-threat landscape, threats from outsourcing, and insider threat. Interestingly, Denmark is one of the few countries that recognize insider threat [18] as a national security threat in their national cybersecurity policy.

In response, the Danish government established the Agency for Digitization as part of the Ministry of Finance, while the Danish police have set up the National Cyber Crime Center (NC3). Specific guidelines targeted at government ICT units and suppliers to the government, are complemented with a commitment to building specific protections for the energy and telecommunications sector, working with international partners, and better explaining cybersecurity implications of the actions of individual citizens.

Our analysis shows that Denmark is one of the safest countries in the world from a cybersecurity perspective.

Denmark	Avg number of attacks per host	Percentage of attacked hosts
2010	3.07	0.40
2011	2.28	0.36

Moreover, despite having a strong record on cybersecurity, Denmark's citizens showed a marked improvement—with an approximately 25% improvement in the number of attacks per host and a 10% improvement in the percentage of attacked hosts.

In terms of the type of threat, Denmark is primarily targeted by Trojans, followed by viruses and worms. Nonetheless, there is room for improvement—our data shows that Danes are more vulnerable to malware based on misleading software than other OECD countries. Denmark should be able to achieve even better results by an education program that clearly brings out the risks posed by misleading software such as fake anti-virus programs and fake disk cleanup utilities (Figs. 5.83, 5.84, 5.85, 5.86, 5.87, 5.88, 5.89, and 5.90).

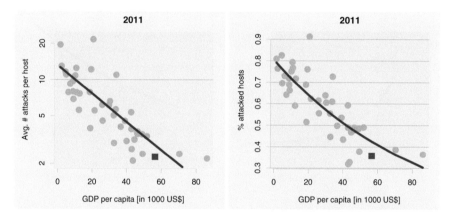

Fig. 5.83 Average number of attacks per host (*left*) and percentage of attacked host (*right*). *Blue line*: predicted values based on GDP-only model

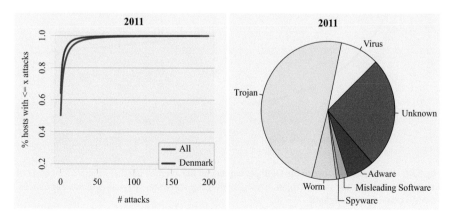

Figs. 5.84 Denmark: Empirical cumulative distribution of % of hosts with less than or equal to x attacks, and **5.85** Denmark: Distribution of attacks by type of malware

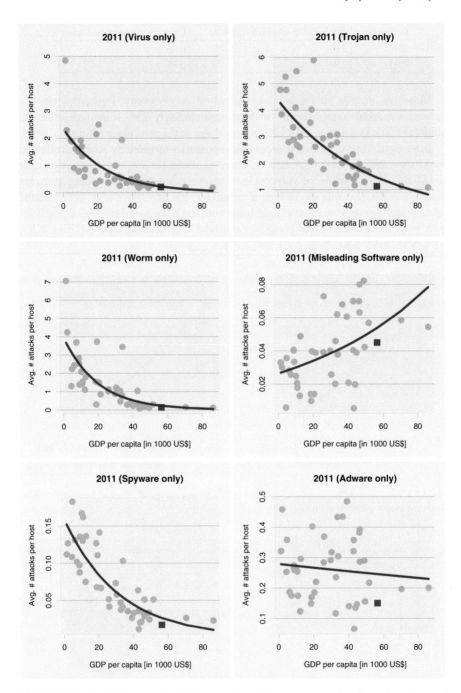

Fig. 5.86 Relationship between the GDP per capita and the average number of attacks on hosts of a country separately for virus, Trojan, worm, misleading software, spyware and adware attacks. Selected countries highlighted

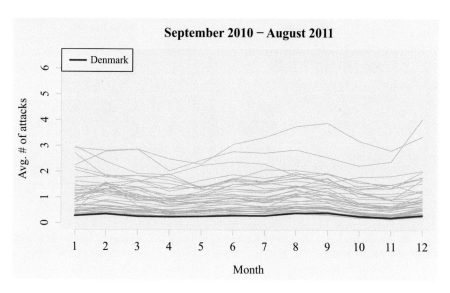

Fig. 5.87 Timeline of average monthly attacks per host by country for the 2011 time period

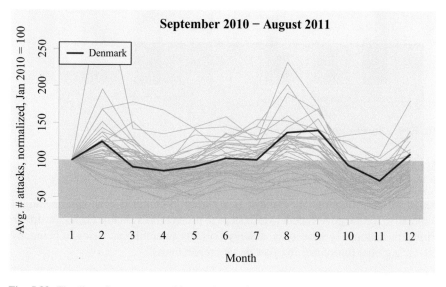

Fig. 5.88 Timeline of average monthly attacks per host by country for the 2011 time period. Normalized by attack count in September 2010

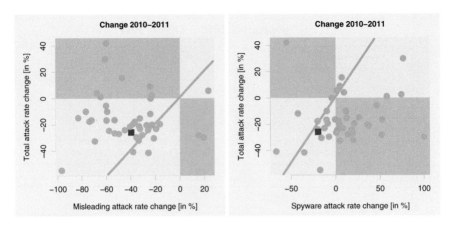

Fig. 5.89 Attack frequency changes of misleading software and spyware in relation to overall change of attack frequency

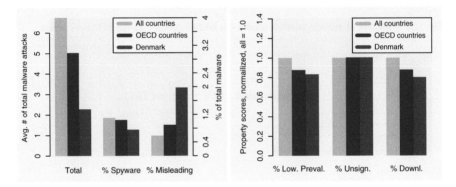

Fig. 5.90 Attack frequencies and host properties

5.12   Finland

As in the case of Belgium, Finland is a highly IT-centered economy, making it imperative for the government to take steps to protect this valuable economic driver. Finland's national cybersecurity strategy [19] has an ambitious vision of being, by 2016, "a global forerunner in cyber threat preparedness".

Operationally, Ministries are in charge of cybersecurity in organizations and in affairs that fall within their legislative scope. In addition, Finland envisages the creation of a Cyber Security Center which operates on a 24/7/365 basis. A heavy investment in cybersecurity research and development, as well as international cooperation is envisaged.

Finland's strategic guidelines for cybersecurity include: (1) creating a collaboration framework within which all stakeholders might collaborate, (2) improve cybersecurity situation awareness and threat prediction, (3) ensure the development of in-house expertise and knowledge for businesses to handle their own cybersecurity, (4) strengthen police knowledge and expertise in handling cybersecurity incidents, (5) ensure that defense forces have the requisite cybersecurity capabilities, (6) enhance awareness of cybersecurity throughout society by improved education and training, and (7) develop appropriate legislation.

Our analysis shows that Finland is one of the most secure countries from a cybersecurity perspective.

Finland	Avg number of attacks per host	Percentage of attacked hosts
2010	3.13	0.41
2011	2.11	0.33

Despite being extremely safe from a cybersecurity perspective in 2010, Finland managed to make significant strides in a positive direction in 2011, making them even safer in 2011.

In terms of the type of threat, Finland is primarily targeted by Trojans, followed closely by viruses and adware. In virtually every area, Finns do much better than their counterparts in countries with a similar GDP. The only place where there is room for some improvement is behavioral. In terms of downloading unsigned and low-prevalence (rare) binaries, Finns seem to act more like their OECD brethren than one would have expected. Perhaps this presents an opportunity for improved education about the risks of downloading low prevalence and unsigned binaries. But all in all, Finland is one of the safest countries in the world from a cybersecurity perspective (Figs. 5.91, 5.92, 5.93, 5.94, 5.95, 5.96, 5.97, and 5.98).

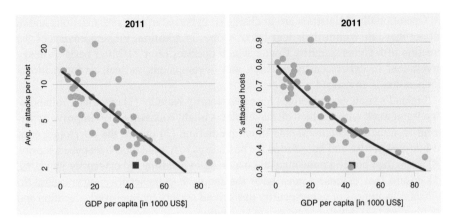

Fig. 5.91 Average number of attacks per host (*left*) and percentage of attacked host (*right*). *Blue line*: predicted values based on GDP-only model

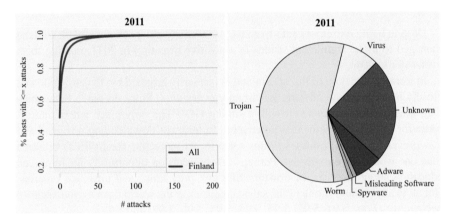

Figs. 5.92 Finland: Empirical cumulative distribution of % of hosts with less than or equal to x attacks, and **5.93** Finland: Distribution of attacks by type of malware

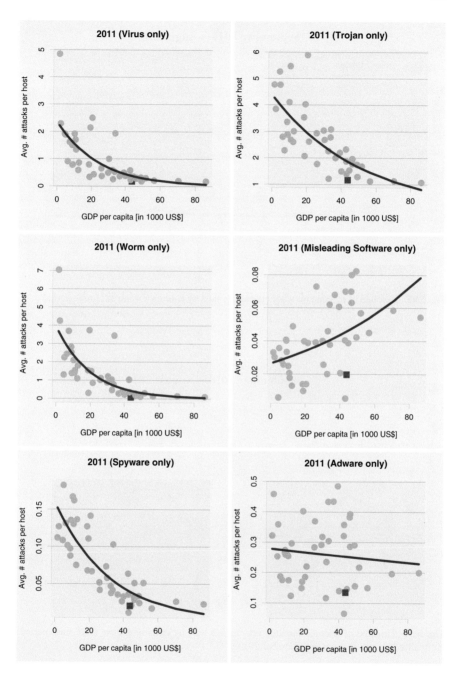

Fig. 5.94 Relationship between the GDP per capita and the average number of attacks on hosts of a country separately for virus, Trojan, worm, misleading software, spyware and adware attacks. Selected countries highlighted

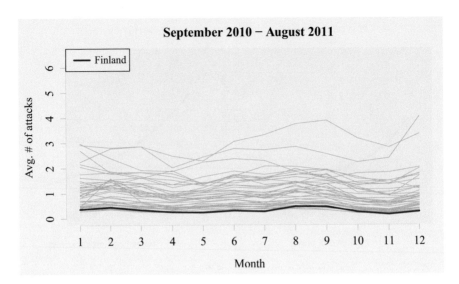

Fig. 5.95 Timeline of average monthly attacks per host by country for the 2011 time period

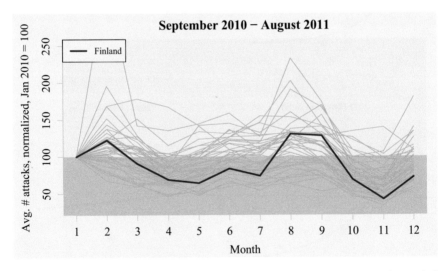

Fig. 5.96 Timeline of average monthly attacks per host by country for the 2011 time period. Normalized by attack count in September 2010

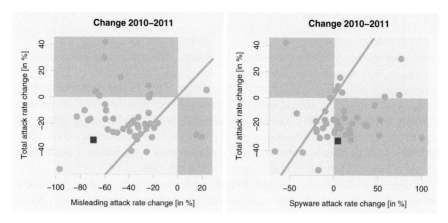

Fig. 5.97 Attack frequency changes of misleading software and spyware in relation to overall change of attack frequency

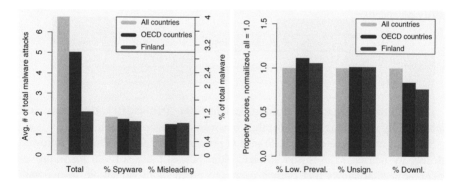

Fig. 5.98 Attack frequencies and host properties

5.13 France

France set up the Agence nationale de la sécurité des systèmes d'information or ANSSI in 2009 in order to counter network-based threats to its security. The French cybersecurity strategy states, amongst other things, that "foreign States or terrorist groups could attack the critical infrastructures of States that they consider as ideologically hostile" [20, p. 11]. They suggest the enhanced use of cryptography as one way to address this issue. The document expresses the need to protect critical infrastructure such as telecom networks and financial data repositories. Protection of private personal information is also considered a key issue.

In order to address these concerns, France's cybersecurity strategy suggests a sevenfold strategy that (1) attempts to anticipate attacks before they occur, (2) detect, analyze and respond to these attacks rapidly when in fact they do occur, (3) enhance the pool of trained personnel with expertise in cybersecurity, e.g. through the creation of a cyber-defense research center, (4) master the use of—and innovate in the creation of—the most advanced cybersecurity tools and techniques, (5) appropriately adapt and enhance French law so as to provide legal remedies to cybersecurity problems, (6) increase international collaborations in view of the transnational nature of the cybersecurity problem, and (7) communicate better with all stakeholders.

Our analysis shows that France has some room for improvement from a cybersecurity perspective.

France	Avg number of attacks per host	Percentage of attacked hosts
2010	6.52	0.63
2011	4.54	0.54

Though there was a marked improvement from 2010 to 2011, there were still substantially more attacks per host than in several advanced economies.

In terms of the type of threat, France is primarily targeted by Trojans, followed closely by worms, viruses and adware. Nonetheless, there is room for improvement—our data shows that French users are much more vulnerable to malware based on misleading software and adware than other nations with comparable GDPs. Moreover, when compared to both OECD and non-OECD countries, French users are more vulnerable to misleading software. France should be able to achieve even better results by an education program that clearly brings out the risks posed by misleading software such as fake anti-virus programs and fake disk cleanup utilities and highlighting the bad impact of adware (Figs. 5.99, 5.100, 5.101, 5.102, 5.103, 5.104, 5.105, and 5.106).

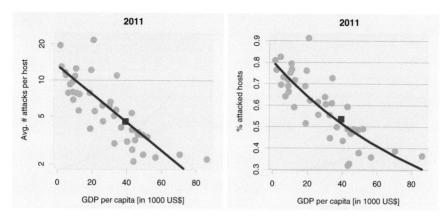

Fig. 5.99 Average number of attacks per host (*left*) and percentage of attacked host (*right*). *Blue line*: predicted values based on GDP-only model

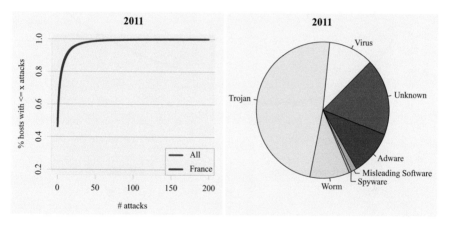

Figs. 5.100 France: Empirical cumulative distribution of % of hosts with less than or equal to x attacks, and **5.101** France: Distribution of attacks by type of malware

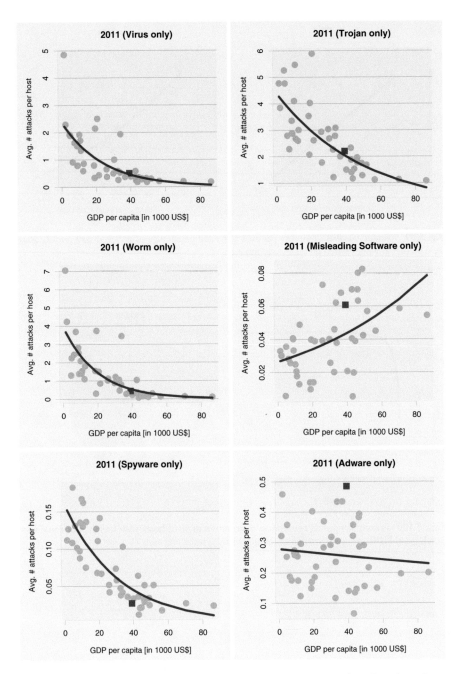

Fig. 5.102 Relationship between the GDP per capita and the average number of attacks on hosts of a country separately for virus, Trojan, worm, misleading software, spyware and adware attacks. Selected countries highlighted

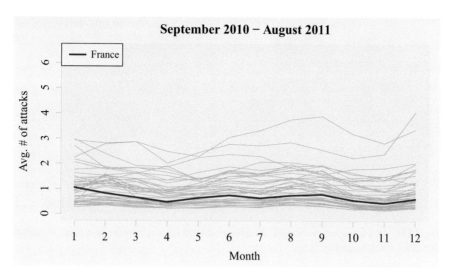

Fig. 5.103 Timeline of average monthly attacks per host by country for the 2011 time period

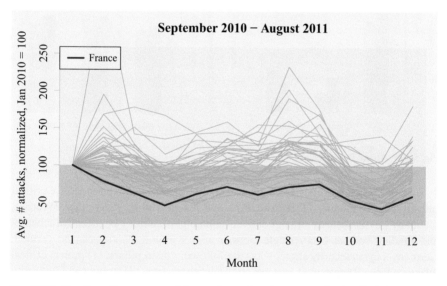

Fig. 5.104 Timeline of average monthly attacks per host by country for the 2011 time period. Normalized by attack count in September 2010

5.14 Germany

Germany's National Cyber Security Strategy [21] is centered around the civilian National Cyber Security Council, along with units in the German Army, police, and security services. The National Cyber Security Council reports to the Federal Office for Information Security.

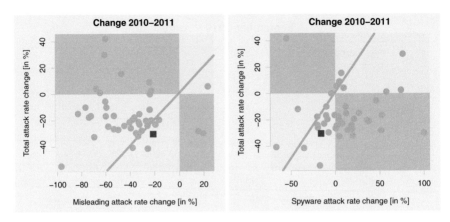

Fig. 5.105 Attack frequency changes of misleading software and spyware in relation to overall change of attack frequency

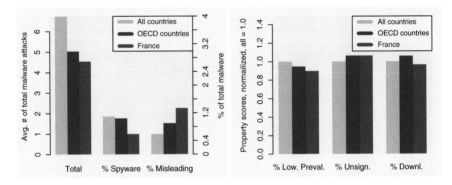

Fig. 5.106 Attack frequencies and host properties

Like its French neighbor, Germany is heavily concerned with protection of critical infrastructure including IT infrastructure and assets. Unlike most other countries, Germany's cybersecurity strategy actively contemplates setting up incentive schemes and funding so that every single German user's online identity can be verified. Germany's cybersecurity strategy has the following main pillars: (1) protect critical infrastructure, (2) protect IT systems and IT infrastructure, (3) strengthen IT security in government, (4) partner with businesses, research organizations, and academia, (5) crackdown on cyber-crime by strengthening law enforcement agencies' capabilities in the cyber domain, (6) foster greater international cooperation, particularly within the EU and NATO, (7) develop and maintain a usable pool of cybersecurity tools and engage in continuous exercises to build capacity in the use of these tools.

Germany's cybersecurity statistics are summarized in the table below.

Germany	Avg number of attacks per host	Percentage of attacked hosts
2010	4.09	0.52
2011	3.09	0.43

Even though Germany did very well in terms of cybersecurity in 2010, there was a significant improvement in cybersecurity in 2011.

Germany is most at risk from Trojans, followed by viruses, worms, and adware. Nevertheless, in terms of virtually every single category of malware, Germany does better than peers with a similar per capita GDP and is one of the safest countries in the world from a cybersecurity perspective.

Nonetheless, there is room for improvement. The percentage of low prevalence and unsigned binaries downloaded by Germans is higher than that in other countries, suggesting that education campaigns that aim to highlight the risks of downloading unsigned and uncommon binaries could further reduce the incidence of malware within Germany (Figs. 5.107, 5.108, 5.109, 5.110, 5.111, 5.112, 5.113, and 5.114).

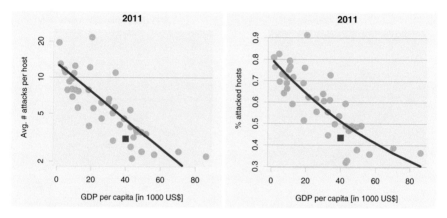

Fig. 5.107 Average number of attacks per host (*left*) and percentage of attacked host (*right*). *Blue line*: predicted values based on GDP-only model

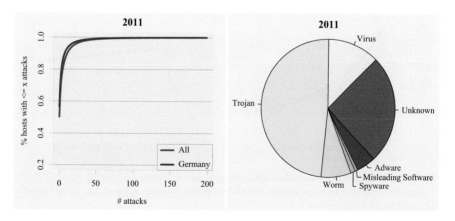

Figs. 5.108 Germany: Empirical cumulative distribution of % of hosts with less than or equal to x attacks, and **5.109** Germany: Distribution of attacks by type of malware

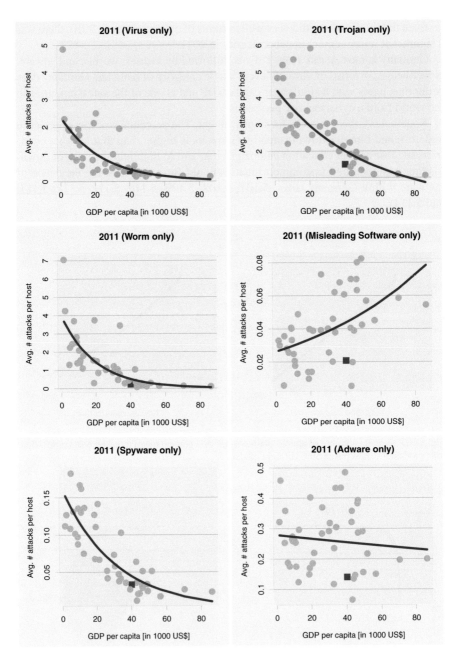

Fig. 5.110 Relationship between the GDP per capita and the average number of attacks on hosts of a country separately for virus, Trojan, worm, misleading software, spyware and adware attacks. Selected countries highlighted

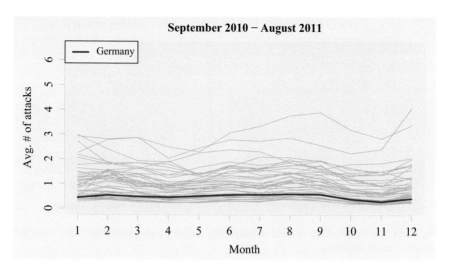

Fig. 5.111 Timeline of average monthly attacks per host by country for the 2011 time period

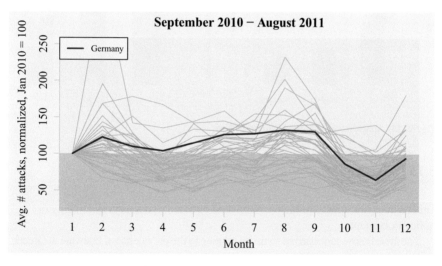

Fig. 5.112 Timeline of average monthly attacks per host by country for the 2011 time period. Normalized by attack count in September 2010

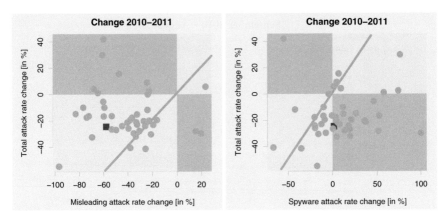

Fig. 5.113 Attack frequency changes of misleading software and spyware in relation to overall change of attack frequency

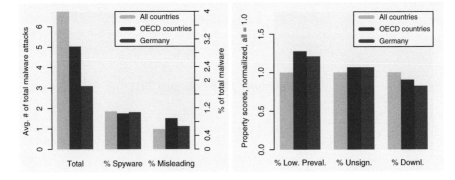

Fig. 5.114 Attack frequencies and host properties

5.15   Greece

Greece does not appear to have a published national cybersecurity strategy or policy document.

The table below summarizes statistics relating to the prevalence of malware in Greece.

Greece	Avg number of attacks per host	Percentage of attacked hosts
2010	7.39	0.65
2011	6.16	0.62

Despite Greece's many problems, the nation was able to make progress from 2010 to 2011, reducing the average number of incidents per host by almost one sixth, though the reduction in the percentage of attacked hosts seems small.

Malware on Greek hosts is dominated by Trojans, followed by worms, viruses, and adware. However, the incidence of these types of malware in Greece is more or less consistent with expectations from countries with a similar GDP. Behaviorally, Greek users' behavior seems in line with that of users from OECD countries (Figs. 5.115, 5.116, 5.117, 5.118, 5.119, 5.120, 5.121, and 5.122).

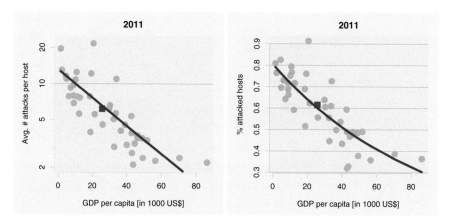

Fig. 5.115 Average number of attacks per host (*left*) and percentage of attacked host (*right*). *Blue line*: predicted values based on GDP-only model

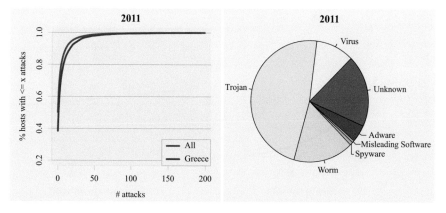

Figs. 5.116 Greece: Empirical cumulative distribution of % of hosts with less than or equal to x attacks, and **5.117** Greece: Distribution of attacks by type of malware

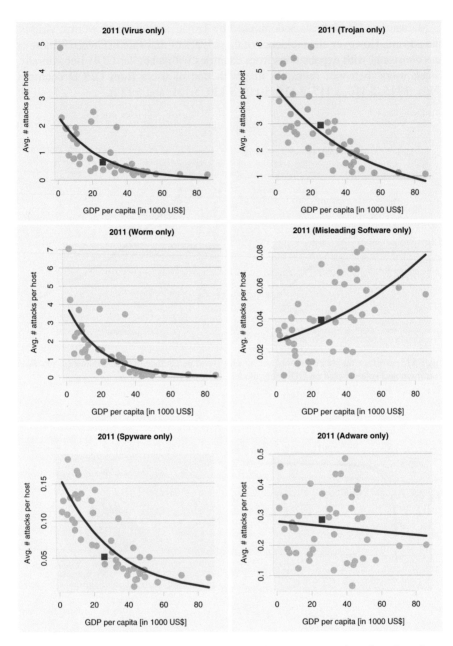

Fig. 5.118 Relationship between the GDP per capita and the average number of attacks on hosts of a country separately for virus, Trojan, worm, misleading software, spyware and adware attacks. Selected countries highlighted

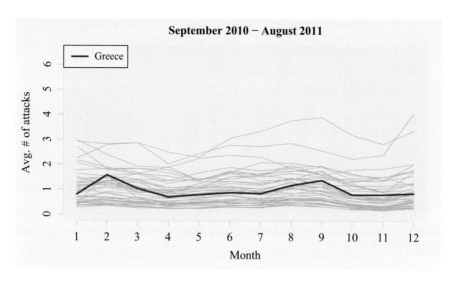

Fig. 5.119 Timeline of average monthly attacks per host by country for the 2011 time period

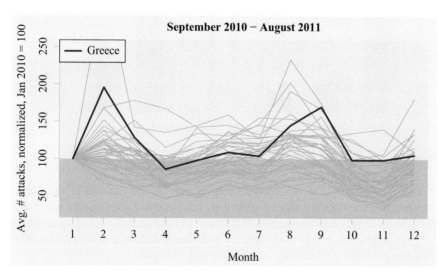

Fig. 5.120 Timeline of average monthly attacks per host by country for the 2011 time period. Normalized by attack count in September 2010

5.16   Hong Kong

Hong Kong's cybersecurity strategy document [22] released in 2011 identifies the Security Bureau and the Office of the Government Chief Information Officer as the parties principally responsible for maintaining a high level of preparedness on the cybersecurity front. The Hong Kong government's principal interests are to

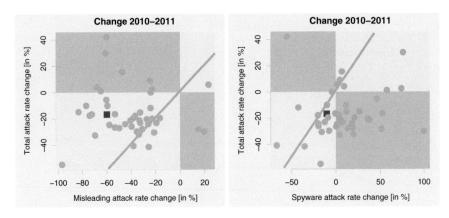

Fig. 5.121 Attack frequency changes of misleading software and spyware in relation to overall change of attack frequency

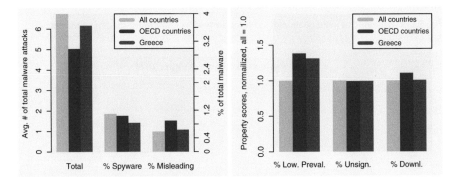

Fig. 5.122 Attack frequencies and host properties

maintain security of critical infrastructure, security of network and IT infrastructure, handle cyber-crime under existing legal instruments rather than new technology-specific legislation, formulate CERTs that carry out periodic exercises to ensure cyber-preparedness, and promote specific web sites and ISO-certification organizations that provide cybersecurity information and education.

Hong Kong has set up specific organizations such as the Internet Infrastructure Liaison Group that brings multiple government and corporate stakeholders to the table. It also participates in cybersecurity exercises run by the Asia Pacific CERT. It maintains an important web site (www.infosec.gov.hk) which provides relevant information to government and business entities.

The table below reports cybersecurity statistics for Hong Kong.

Hong Kong	Avg number of attacks per host	Percentage of attacked hosts
2010	6.91	0.63
2011	5.05	0.55

We see that Hong Kong experienced a huge improvement in cybersecurity from 2010 to 2011, reducing the average number of attacks per host by almost 27% and also achieving impressive reductions in the percentage of attacked hosts. On average, Hong Kong defends against different types of malware as well or better than countries with a similar GDP.

Nonetheless, there is some room for improvement. Our analysis of user behavior in Hong Kong suggests that users tend to download more binaries (in all) as well as more low prevalence binaries than those in comparable countries. This suggests that an education campaign highlighting the risks of downloading binaries, especially those that are not well-known, could be beneficial to Hong Kong.

In closing, we note that we treat Hong Kong separately from China in our study because the WINE data distinguished between the two, rather than for any other reason (Figs. 5.123, 5.124, 5.125, 5.126, 5.127, 5.128, 5.129, and 5.130).

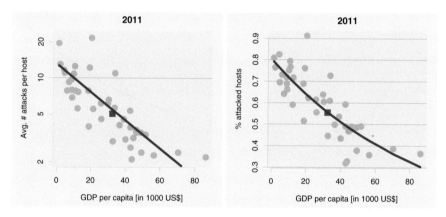

Fig. 5.123 Average number of attacks per host (*left*) and percentage of attacked host (*right*). *Blue line*: predicted values based on GDP-only model

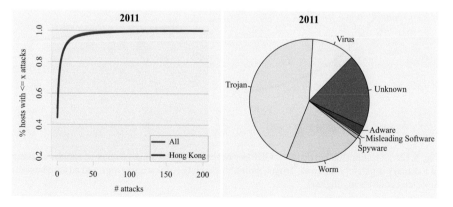

Figs. 5.124 Hong Kong: Empirical cumulative distribution of % of hosts with less than or equal to x attacks, and **5.125** Hong Kong: Distribution of attacks by type of malware

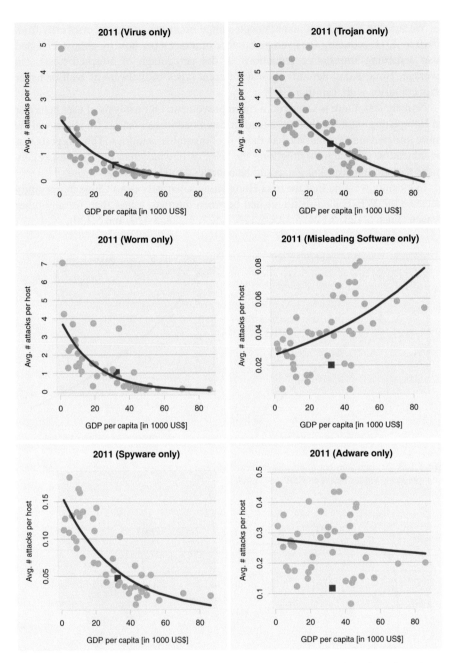

Fig. 5.126 Relationship between the GDP per capita and the average number of attacks on hosts of a country separately for virus, Trojan, worm, misleading software, spyware and adware attacks. Selected countries highlighted

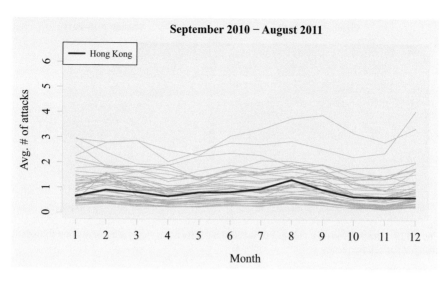

Fig. 5.127 Timeline of average monthly attacks per host by country for the 2011 time period

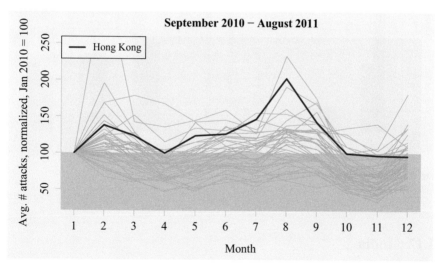

Fig. 5.128 Timeline of average monthly attacks per host by country for the 2011 time period. Normalized by attack count in September 2010

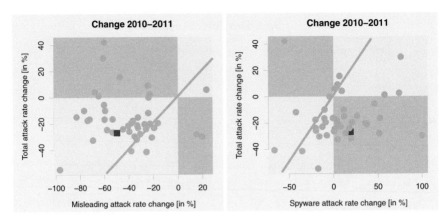

Fig. 5.129 Attack frequency changes of misleading software and spyware in relation to overall change of attack frequency

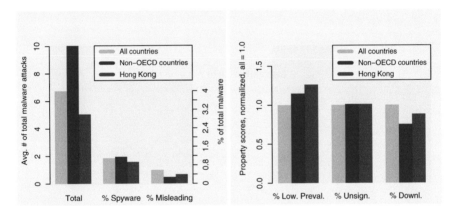

Fig. 5.130 Attack frequencies and host properties

5.17 India

India released its National Cyber Security Policy in 2013 [23]. Because of the fact that IT is a driver of significant growth in India, any cyber vulnerability of the Indian IT sector could be devastating to Indian growth. For instance, Indian IT companies like Tata Consultancy Services, Infosys, and Wipro are behemoths that provide a host of data management, enterprise, network, and other ICT services to Fortune 500 and other large corporations worldwide. A cyber-attack on data centers housed in India would have wide-ranging effects on India's economy.

Because of these factors, India's cybersecurity strategy calls for the creation of 500,000 highly skilled cybersecurity experts within 5 years (i.e. by 2018). This is a very ambitious goal by any reckoning. Nonetheless, India's government is taking steps to make this a reality, even if the schedule slips by some years.

India's cybersecurity strategy calls for the provision of financial incentives to companies to make better investments in cybersecurity, the creation of Chief Information Security Officer (CISO) roles within companies, and development of research centers and training centers to increase capacity in cyber security. India could potentially be a world leader in cybersecurity—with over 20 million trained computer scientists, the country has a rich group of professionals to draw upon in building and offering cybersecurity services to other countries.

Nevertheless, this optimistic outlook for the future must be tempered by the cold, hard facts summarized below.

India	Avg number of attacks per host	Percentage of attacked hosts
2010	17.93	0.83
2011	19.53	0.81

Unlike many other countries, India saw an *increase* in the average number of attacks per host when going from 2010 to 2011, with the percentage of attacked hosts staying more or less static. The numbers suggest that a little less than 20% of Indians practice good cyber hygiene and do not end up being attacked. However, the remaining 80% of hosts are heavily attacked. There is no getting around the fact that India (along with South Korea) compete for the dubious distinction of being the most cyber-vulnerable nations in the world.

Hosts in India are primarily targeted by worms, followed by Trojans, and viruses, which is different from more advanced economies that tend to be more heavily targeted by Trojans. Indian users also download more binaries and more low-prevalence binaries than their counterparts in other countries. This suggests that educational efforts that highlight the risks of downloading binaries from the Internet may help reduce attacks in India (Figs. 5.131, 5.132, 5.133, 5.134, 5.135, 5.136, 5.137, and 5.138).

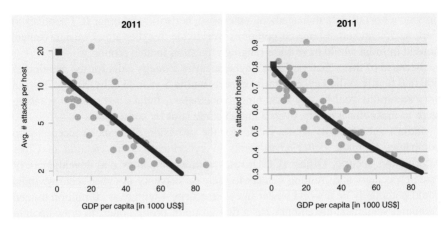

Fig. 5.131 Average number of attacks per host (*left*) and percentage of attacked host (*right*). *Blue line*: predicted values based on GDP-only model

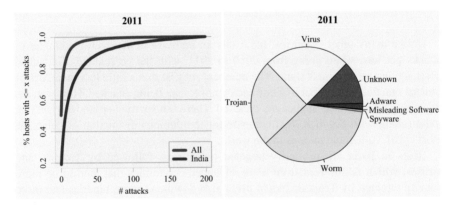

Figs. 5.132 India: Empirical cumulative distribution of % of hosts with less than or equal to x attacks, and **5.133** India: Distribution of attacks by type of malware

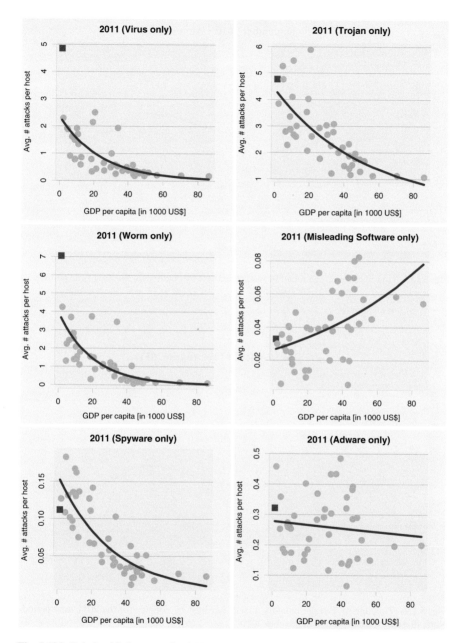

Fig. 5.134 Relationship between the GDP per capita and the average number of attacks on hosts of a country separately for virus, Trojan, worm, misleading software, spyware and adware attacks. Selected countries highlighted

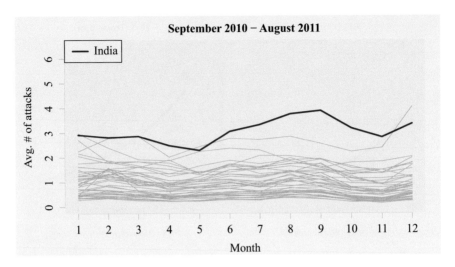

Fig. 5.135 Timeline of average monthly attacks per host by country for the 2011 time period

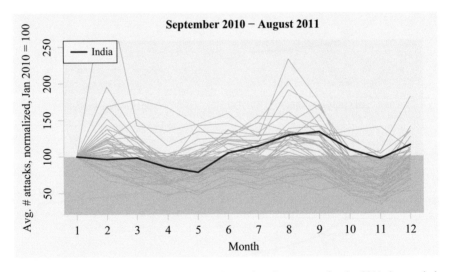

Fig. 5.136 Timeline of average monthly attacks per host by country for the 2011 time period. Normalized by attack count in September 2010

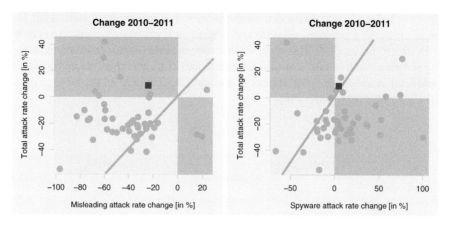

Fig. 5.137 Attack frequency changes of misleading software and spyware in relation to overall change of attack frequency

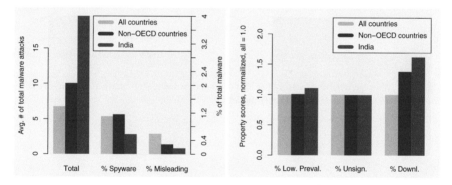

Fig. 5.138 Attack frequencies and host properties

5.18 Ireland

We were unable to find a national cybersecurity strategy for Ireland.

Statistics about attacks on Irish hosts are summarized in the table below.

Ireland	Avg number of attacks per host	Percentage of attacked hosts
2010	5.03	0.57
2011	3.53	0.49

Nevertheless, Ireland does very well from a cybersecurity perspective. The average number of attacks per host is better than several advanced economies (e.g. France). Like most advanced economies, Trojans are the most prevalent cybersecurity threat to Ireland, followed by viruses and worms. But interestingly, Ireland has far more misleading software, spyware and adware on hosts when compared to countries with a similar GDP. This suggests that educational efforts focused on misleading software (such as fake anti-virus programs and disk cleanup programs) could be beneficial in reducing cyber attacks in Ireland (Figs. 5.139, 5.140, 5.141, 5.142, 5.143, 5.144, 5.145, and 5.146).

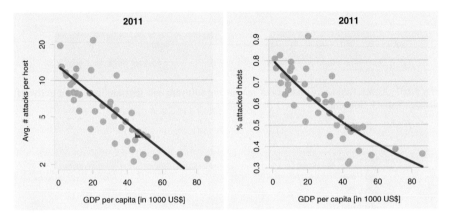

Fig. 5.139 Average number of attacks per host (*left*) and percentage of attacked host (*right*). *Blue line*: predicted values based on GDP-only model

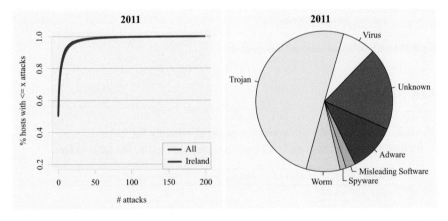

Figs. 5.140 Ireland: Empirical cumulative distribution of % of hosts with less than or equal to x attacks, and **5.141** Ireland: Distribution of attacks by type of malware

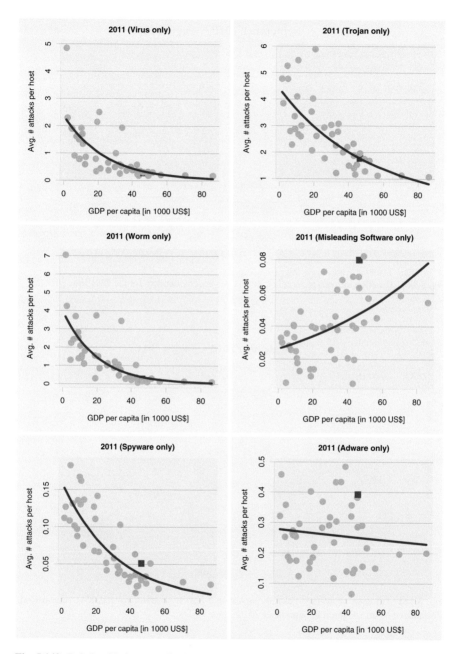

Fig. 5.142 Relationship between the GDP per capita and the average number of attacks on hosts of a country separately for virus, Trojan, worm, misleading software, spyware and adware attacks. Selected countries highlighted

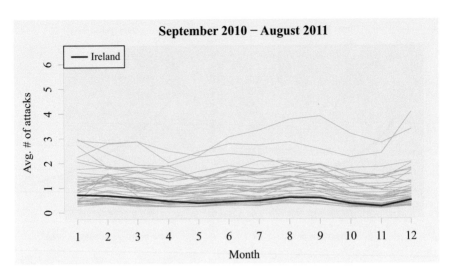

Fig. 5.143 Timeline of average monthly attacks per host by country for the 2011 time period

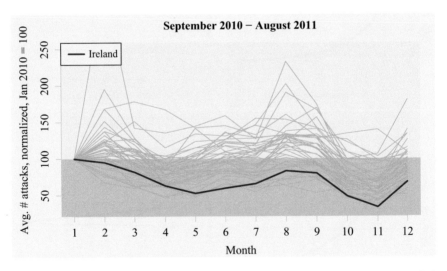

Fig. 5.144 Timeline of average monthly attacks per host by country for the 2011 time period. Normalized by attack count in September 2010

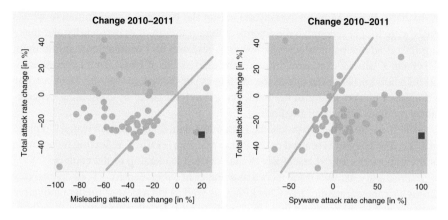

Fig. 5.145 Attack frequency changes of misleading software and spyware in relation to overall change of attack frequency

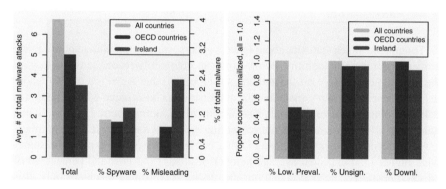

Fig. 5.146 Attack frequencies and host properties

5.19 Israel

The State of Israel has long had a very strong reputation as a global leader in cybersecurity. Going as far back as the 1990s, Israeli military officials set up cybersecurity cells within diverse units. For instance, [24, 25] states that the Israeli Security Agency (*Shabaq*) had set up a cybersecurity group well before the country set up its national Critical Infrastructure Policy in December 2002. Later, the National Information Security Agency (*Re 'em*) grew out of this ISA group.

In 2010, Prime Minister Netanyahu set up the National Cyber Initiative whose goals were to: (1) recognize the interdisciplinary nature of cybersecurity and make significant advances in education, training, research and development in the field, (2) create a "shield" that would protect the entire state of Israel from cyber-attacks, (3) help build up Israel as one of the top five cyber-powers in the world and become a dominant market leader in the field, and (4) defend Israel by combining cybersecurity methods with other methods.

With these next steps in place, the Israeli National Cyber Bureau was set in place in 2011. The Bureau is housed in the Prime Minister's Office, reflecting recognition of the importance of cybersecurity to the very top leadership of the country.

Interestingly, Israel's track record as a leader in cybersecurity—both defensively through strong companies like CheckPoint as well as via a slew of highly innovative startups, makes it surprising that attack statistics for Israel are not better than those shown below.

Israel	Avg number of attacks per host	Percentage of attacked hosts
2010	7.49	0.67
2011	6.61	0.61

Though the attack statistics about Israel's cybersecurity capabilities show a fairly moderate defensive capability, two factors can potentially reconcile our data with the general perception (and the authors' strong belief) that Israel is a market leader in cybersecurity (an assessment that the authors of this book agree with). The first is that a very large number of people have a particularly intense hatred for Israel—the number of attempted attacks is not something we could measure. If those numbers are significantly higher than those in other countries, they may explain the somewhat higher than expected average number of attacks per host and percentage of attacked hosts. The second is that our data looks at consumer machines only. It is possible that the high level of cybersecurity expertise and awareness, at the very top echelons of Israeli academic, government, and industry, is not reflected in the "man on the street" who may take a more cavalier attitude towards cybersecurity. Both these observations might explain why, for Israel, our data shows Israel's performance to be more or less consistent with what one would expect from a country with Israel's per-capita GDP.

Israel primarily suffers from Trojan attacks, followed by worms and viruses. Interestingly, our data shows that Israelis exhibit less insecure behavior (downloading binaries or low prevalence or unsigned binaries) than their counterparts—both in all other countries as well as in more advanced OECD countries (Figs. 5.147, 5.148, 5.149, 5.150, 5.151, 5.152, 5.153, and 5.154).

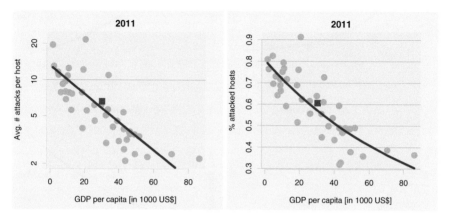

Fig. 5.147 Average number of attacks per host (*left*) and percentage of attacked host (*right*). *Blue line*: predicted values based on GDP-only model

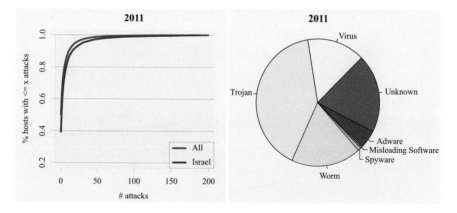

Figs. 5.148 Israel: Empirical cumulative distribution of % of hosts with less than or equal to x attacks, and **5.149** Israel: Distribution of attacks by type of malware

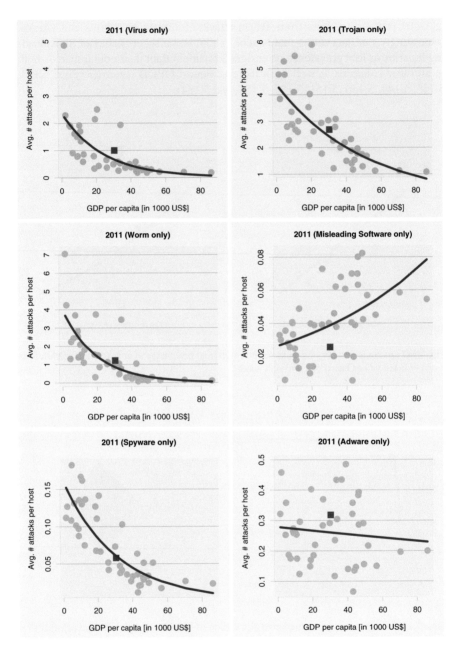

Fig. 5.150 Relationship between the GDP per capita and the average number of attacks on hosts of a country separately for virus, Trojan, worm, misleading software, spyware and adware attacks. Selected countries highlighted

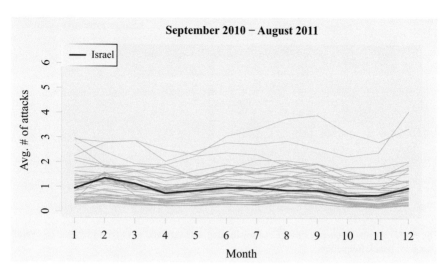

Fig. 5.151 Timeline of average monthly attacks per host by country for the 2011 time period

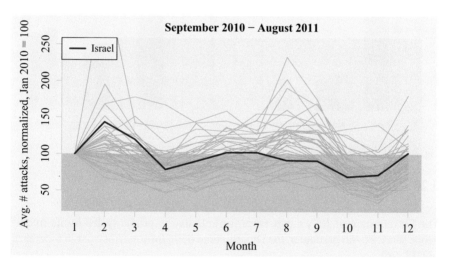

Fig. 5.152 Timeline of average monthly attacks per host by country for the 2011 time period. Normalized by attack count in September 2010

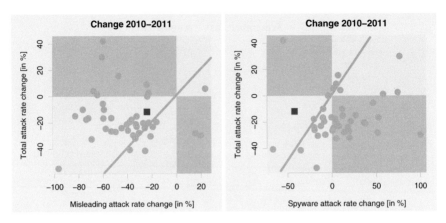

Fig. 5.153 Attack frequency changes of misleading software and spyware in relation to overall change of attack frequency

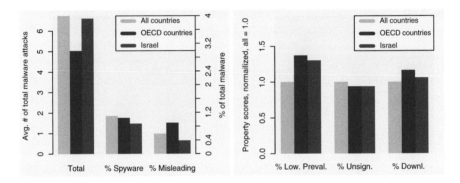

Fig. 5.154 Attack frequencies and host properties

5.20 Italy

The Government of Italy's cybersecurity initiative is primarily led by its intelligence services—in particular, the Dipartimento delle Informazione per la Sicurezza or DIS [26].

Italy's cybersecurity strategy revolves around six strategic guidelines that include (1) capacity development in cybersecurity, (2) critical infrastructure protection, (3) developing public private partnerships focused on cybersecurity, (4) helping shape a national culture of cybersecurity through academic and others types of public awareness campaigns, (5) effective counteraction of cyber-crime, and (6) participation in appropriate international cybersecurity related cooperative ventures.

In order to achieve this, Italy has created the National Information Security (NIS) Authority which will spearhead cooperation with international entities, especially those within the EU and NATO. Moreover, this organization also is charged with running a variety of cyber defense exercises, both within Italy as well as in cross-border EU settings. Interestingly, the Italian policy document specifies a goal of

developing "national dissuasion and deterrence capabilities in cyber space" though
it is vague on what these deterrents might be or who might need to be deterred.

Italy's situation with respect to cyber-vulnerability is summarized in the table below.

Italy	Avg number of attacks per host	Percentage of attacked hosts
2010	8.03	0.71
2011	5.66	0.61

The results show a sharp improvement in Italy's situation, reflecting an over 30%
improvement on the average number of attacks per host from 2010 to 2011.

Italy's major cybersecurity threats are Trojans, worms and viruses in that order.
The prevalence of adware and misleading software in Italy is far above the norm for
countries with a similar GDP, suggesting that education about the risks of misleading software such as fake anti-viruses and fake disk cleanup utilities could help
improve the level of cyber vulnerability in general (Figs. 5.155, 5.156, 5.157, 5.158,
5.159, 5.160, 5.161, and 5.162).

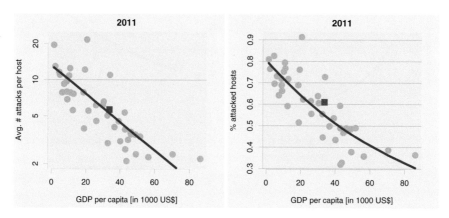

Fig. 5.155   Average number of attacks per host (*left*) and percentage of attacked host (*right*). *Blue
line*: predicted values based on GDP-only model

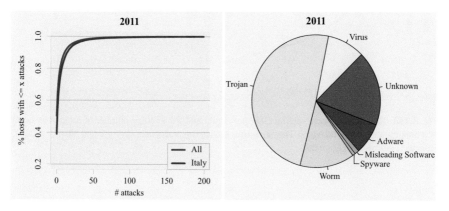

Figs. 5.156   Italy: Empirical cumulative distribution of % of hosts with less than or equal to x
attacks, and **5.157** Italy: Distribution of attacks by type of malware

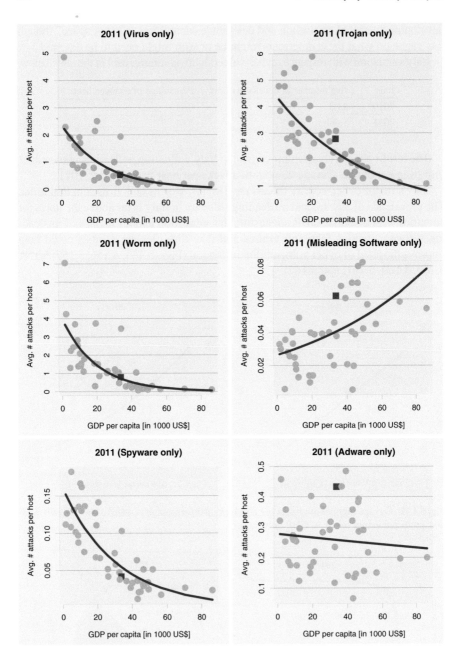

Fig. 5.158 Relationship between the GDP per capita and the average number of attacks on hosts of a country separately for virus, Trojan, worm, misleading software, spyware and adware attacks. Selected countries highlighted

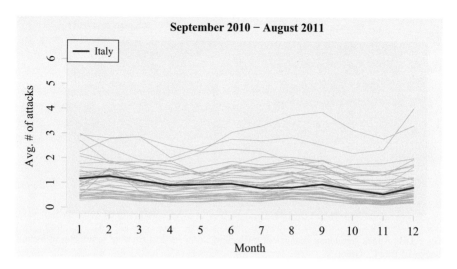

Fig. 5.159 Timeline of average monthly attacks per host by country for the 2011 time period

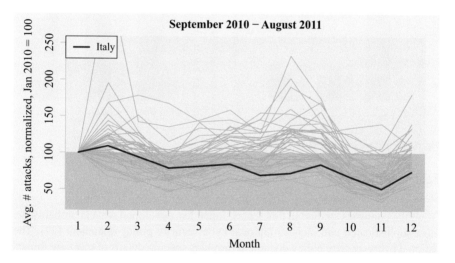

Fig. 5.160 Timeline of average monthly attacks per host by country for the 2011 time period. Normalized by attack count in September 2010

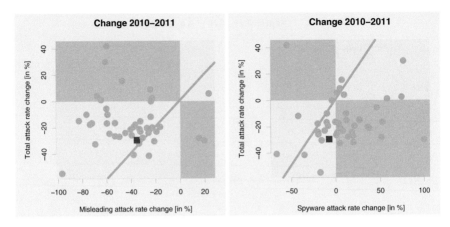

Fig. 5.161 Attack frequency changes of misleading software and spyware in relation to overall change of attack frequency

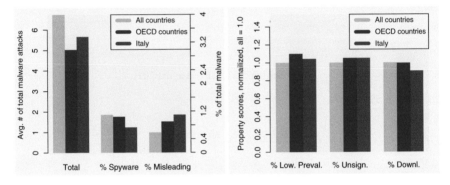

Fig. 5.162 Attack frequencies and host properties

5.21   Japan

Of all the countries studied in this report, Japan has developed what is probably the most sophisticated and incisive national cybersecurity policy document [27]. The national cybersecurity policy document provides specific details of malware threats of concern to Japan including threats to intelligent transportation systems and threats to the Internet of thing. It is the only national cybersecurity strategy document we saw that pointed out the possibility of attacks on devices that are not directly connected to the Internet (e.g. through "air-gapped" attacks [27, 28]).

In order to address these types of threats and to become a world leader in cyber-security, Japan created the National Information Security Center (NISC) as far back as 2005, as well as the Information Security Policy Council (ISPC). Their National Cyber Security Master Plan consists of elements that seek to: (1) develop early warning detection systems for cyber-attacks and appropriate mitigation systems, (2) develop specific methods to protect infrastructure and facilities, (3) develop world class, resilient, cybersecurity platforms, (4) develop cybersecurity deterrence techniques, (5) strengthen international cooperation in cybersecurity, and (6) increase the importance of cybersecurity within corporations.

In order to achieve these goals, Japan recognizes that multiple stakeholders all need to take important steps. Specifically, they identify the role of the government, role of critical infrastructure providers, role of private companies, universities, and research institutions, the role of various types of network operators, as well as the roles of businesses and individual users. In particular, the Japanese cybersecurity strategy document clearly recognizes that smartphones and other smart devices may have huge security holes. They recognize that cybersecurity is a shared responsibility—and that users are responsible not only for their own cybersecurity, but share a collective responsibility.

The table below shows Japan's cybersecurity statistics.

Japan	Avg number of attacks per host	Percentage of attacked hosts
2010	5.84	0.42
2011	2.64	0.32

Japan's already strong cybersecurity in 2010 was improved markedly in 2011. In fact, the number of attacks per host dropped by more than 50%, making Japan's progress truly impressive. In the same vein, the percentage of attacked hosts dropped by almost 25%—another impressive achievement. Simply put, Japan has shown both a strong will and a strong capability of improving cyber defence.

Host machines in Japan are targeted (in descending order of frequency) by Trojans, worms, and viruses. But on every type of malware that we studied, Japanese hosts were better protected than other countries with similar GDPs. Moreover, on virtually every type of "unhygienic" behavior (from a cybersecurity perspective), Japanese users did as well or better than their counterparts in other countries (including OECD countries) (Figs. 5.163, 5.164, 5.165, 5.166, 5.167, 5.168, 5.169, and 5.170).

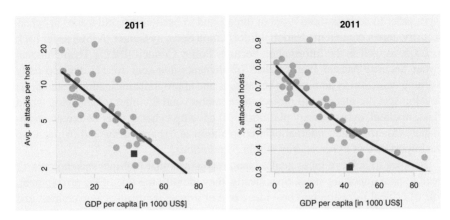

Fig. 5.163 Average number of attacks per host (*left*) and percentage of attacked host (*right*). *Blue line*: predicted values based on GDP-only model

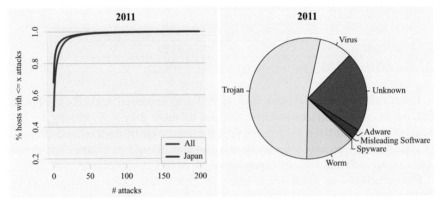

Figs. 5.164 Japan: Empirical cumulative distribution of % of hosts with less than or equal to x attacks, and **5.165** Japan: Distribution of attacks by type of malware

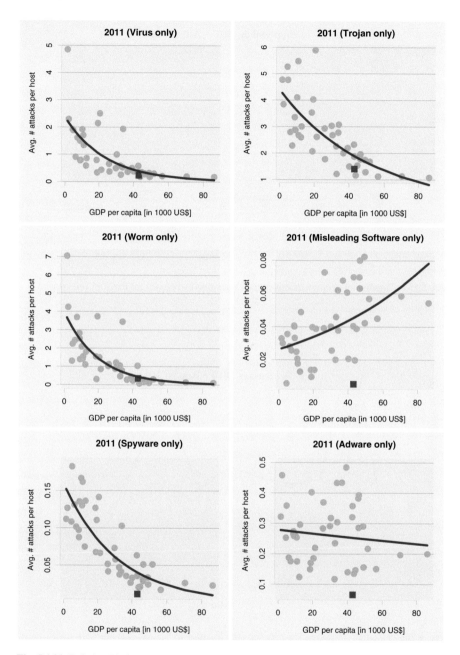

Fig. 5.166 Relationship between the GDP per capita and the average number of attacks on hosts of a country separately for virus, Trojan, worm, misleading software, spyware and adware attacks. Selected countries highlighted

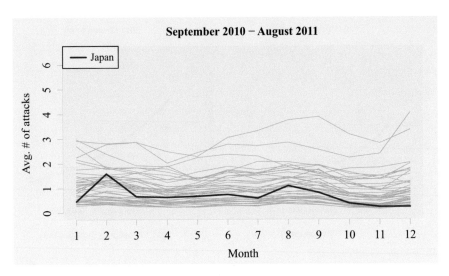

Fig. 5.167 Timeline of average monthly attacks per host by country for the 2011 time period

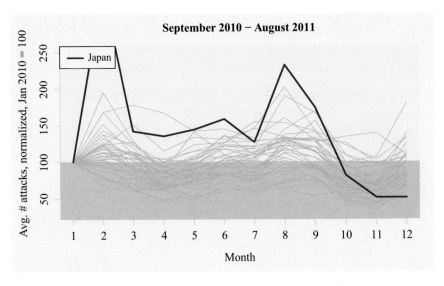

Fig. 5.168 Timeline of average monthly attacks per host by country for the 2011 time period. Normalized by attack count in September 2010

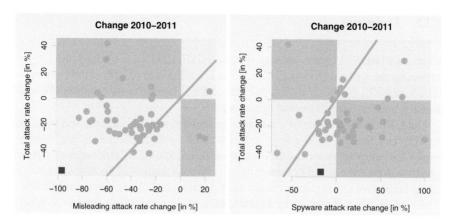

Fig. 5.169 Attack frequency changes of misleading software and spyware in relation to overall change of attack frequency

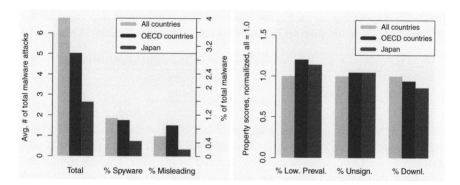

Fig. 5.170 Attack frequencies and host properties

5.22 Malaysia

Malaysia's National Cyber Security Policy (NCSP) [29, 30] is focused primarily at protecting the Critical National Information Infrastructure (CNII) of the country— but additionally, it sets up a chain of command and hierarchy that enables efficient decision making during times of cyber crisis.

The CNII itself consists of the following sectors: national defense and security, banking and finance, ICT, energy, transportation, water, health services, government, emergency services, and food and agriculture. The NCSP envisages an 8-point plan to address cybersecurity threats based on a combination of good governance, a strong legislative and regulatory framework, a strong technical cybersecurity framework, enhancing and promoting a culture of security and capacity building, cybersecurity R&D, standardization and compliance across all parts of the CNII, a strong set of national CERTs, and international cooperation including involvement in cybersecurity forums of the Organization of Islamic Countries (OICs), participation in international bodies, and bilateral agreements with different CERTs worldwide.

Malaysia's cybersecurity structure is coordinated by the National IT Council chaired by the Prime Minister, who receives direct input from the National Cybersecurity Advisory Committee, who in turn receive input from the National Cybersecurity Coordination Committee. Multiple National Cybersecurity Working groups are responsible for each of the 8-points in the 8-point plan referenced above.

The table below summarizes the statistics related to cybersecurity in Malaysia.

Malaysia	Avg number of attacks per host	Percentage of attacked hosts
2010	12.56	0.76
2011	9.68	0.72

Though there was a significant improvement in the average number of attacks per host in going from 2010 to 2011, Malaysia still faces a heavy cyber threat, with about 75% of all Malaysian hosts being attacked by malware. Trojans, worms, and viruses (in that order) constitute the dominate threat vectors. In comparison with other countries with a similar GDP, Malaysia experiences more attacks due to misleading software than their GDP might have led one to expect. In addition, the percentage of both low-prevalence binaries and downloaded binaries on Malaysian hosts is larger than that in comparable countries. These observations suggest that Malaysian citizens need to be better educated about these specific threat vectors. Education programs focused on helping Malaysians identify misleading software such as fake anti-virus utilities and fake disk cleanup utilities (and similar software) could help reduce the incidence of malware in Malaysia. Similarly, enhancing cyber-hygiene by educating Malaysians about the risks of downloading low-prevalence binaries in particular may further help reduce the threat (Figs. 5.171, 5.172, 5.173, 5.174, 5.175, 5.176, 5.177, and 5.178).

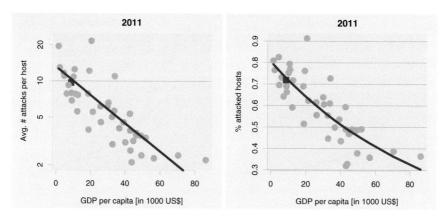

Fig. 5.171   Average number of attacks per host (*left*) and percentage of attacked host (*right*). *Blue line*: predicted values based on GDP-only model

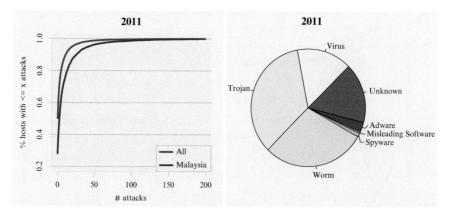

Figs. 5.172   Malaysia: Empirical cumulative distribution of % of hosts with less than or equal to x attacks, and **5.173** Malaysia: Distribution of attacks by type of malware

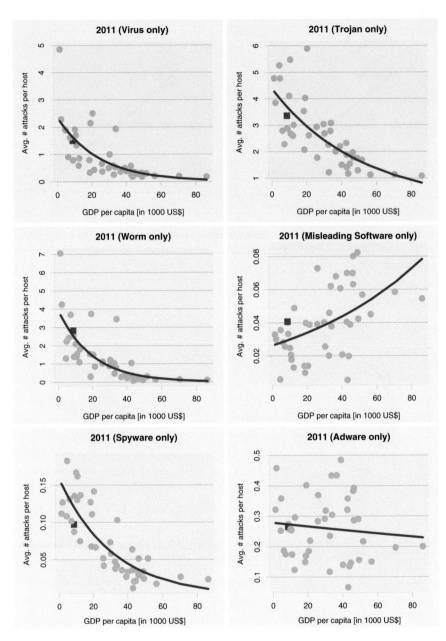

Fig. 5.174 Relationship between the GDP per capita and the average number of attacks on hosts of a country separately for virus, Trojan, worm, misleading software, spyware and adware attacks. Selected countries highlighted

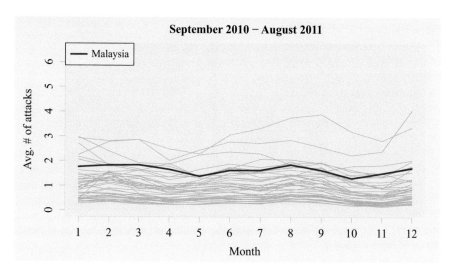

Fig. 5.175 Timeline of average monthly attacks per host by country for the 2011 time period

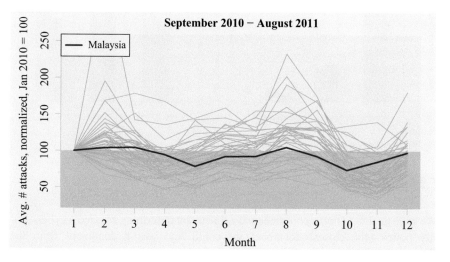

Fig. 5.176 Timeline of average monthly attacks per host by country for the 2011 time period. Normalized by attack count in September 2010

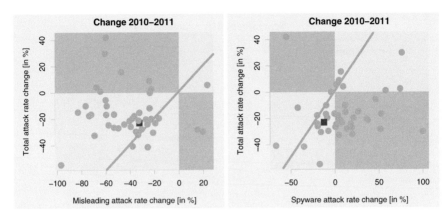

Fig. 5.177 Attack frequency changes of misleading software and spyware in relation to overall change of attack frequency

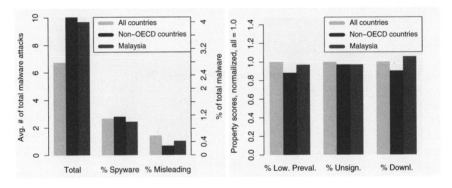

Fig. 5.178 Attack frequencies and host properties

5.23 Mexico

We were unable to find a comprehensive document describing Mexico's cybersecurity policy—the closest document we could find was a document from the Organization of American States (OAS) laying out a cybersecurity position statement common to all OAS member nations [31].

Mexico	Avg number of attacks per host	Percentage of attacked hosts
2010	9.88	0.75
2011	7.96	0.69

Though Mexico did succeed in reducing the average number of attacks per host by almost 20% within 1 year, almost 70% of Mexican hosts continue to be attacked

by malware. Worms and Trojans vie closely for the top spot (in terms of threat vectors) with viruses being the next most common form of malware. In contrast to
some countries, Mexico's malware attacks are consistent with their per capita GDP
(Figs. 5.179, 5.180, 5.181, 5.182, 5.183, 5.184, 5.185, and 5.186).

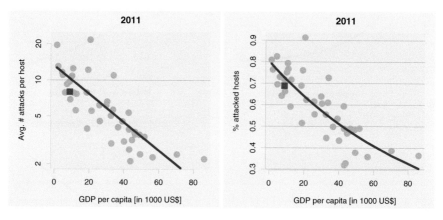

Fig. 5.179   Average number of attacks per host (*left*) and percentage of attacked host (*right*). *Blue
line*: predicted values based on GDP-only model

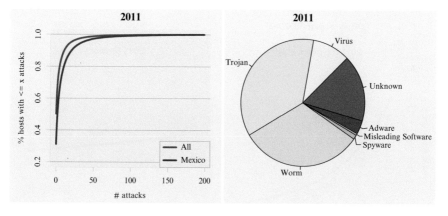

Figs. 5.180   Mexico: Empirical cumulative distribution of % of hosts with less than or equal to x
attacks, and **5.181** Mexico: Distribution of attacks by type of malware

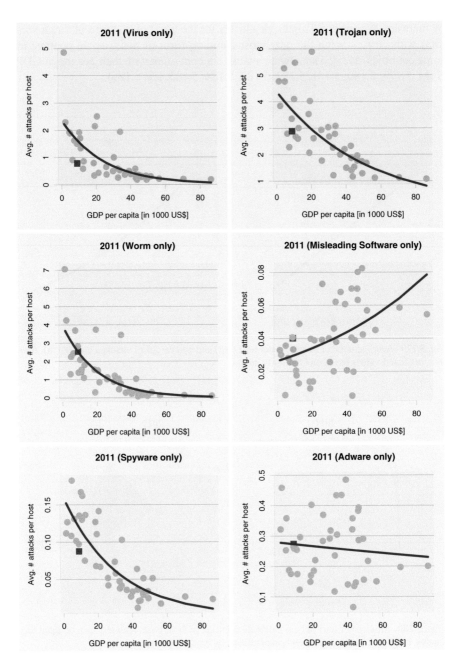

Fig. 5.182 Relationship between the GDP per capita and the average number of attacks on hosts of a country separately for virus, Trojan, worm, misleading software, spyware and adware attacks. Selected countries highlighted

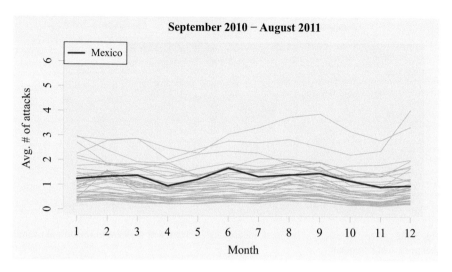

Fig. 5.183 Timeline of average monthly attacks per host by country for the 2011 time period

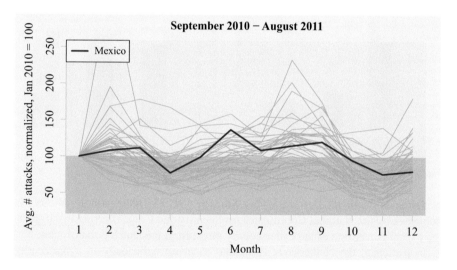

Fig. 5.184 Timeline of average monthly attacks per host by country for the 2011 time period. Normalized by attack count in September 2010

5.24 Netherlands

The Netherlands published its first National Cyber Security Strategy (NCSS1) in 2011. A second such strategy was published in 2014—this strategy focuses on how the Netherlands will handle cybersecurity issues in the 2014–2016 period. The report asserts that 94% of all Dutch households are on the Internet, making them the most connected nation on earth. Moreover, the report asserts that 18% of all Internet traffic flows through the Netherlands. A separate document focuses on the

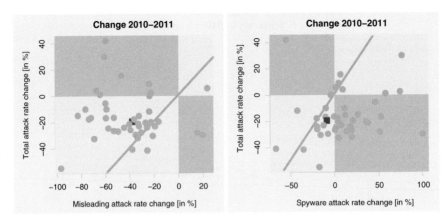

Fig. 5.185 Attack frequency changes of misleading software and spyware in relation to overall change of attack frequency

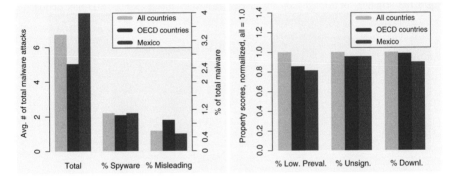

Fig. 5.186 Attack frequencies and host properties

Netherlands' National Defense Cyber Strategy [32, 33]. Interestingly, NCSS2 is published by the National Coordinator for Security and Counterterrorism, suggesting a clear perception of the cybersecurity threat from the Dutch point of view.

In particular, the Netherlands is anxious to invest sufficiently in cybersecurity research, development, and capacity building in order to safeguard against attacks and protect their economy, handling cyber-crime, invest in the development of the right products, and builds and/or participates in the right international cybersecurity partnerships. In order to achieve these objectives, the Netherlands is taking several active steps including analyzing cybersecurity risk, taking more active approaches to mitigate cyberespionage against Dutch citizens and companies, establish whether it makes sense to have a separate (more secure) network devoted to the public institutions and a separate private network, enhancing cooperation between the Dutch military and industry, strengthening the Dutch National Cyber Security Center and Dutch CERTs, building international partnerships and diplomacy, and enhancing education in cybersecurity and investing/supporting cyber innovation, both in companies and universities.

The following table provides a quick summary of the attacks on Dutch hosts.

Netherlands	Avg number of attacks per host	Percentage of attacked hosts
2010	4.36	0.52
2011	3.70	0.49

From a cybersecurity perspective, the Netherlands is one of the safest nations on earth. Despite an already low number of attacks per host, the Dutch were able to reduce this by over 15% in just 1 year.

The overwhelming malware category present on Dutch hosts is Trojans, followed by viruses and worms in that order. In fact, the percentage of Trojans on Dutch hosts (over 50%) is higher than in most other countries. Moreover, the rate of misleading software in Dutch hosts is much higher than in other OECD countries, suggesting that this is a common way for Trojans to be injected into Dutch hosts. We believe that improving education on how misleading software should be handled by Dutch consumers would reduce the overall malware attack rate on Dutch hosts (Figs. 5.187, 5.188, 5.189, 5.190, 5.191, 5.192, 5.193, and 5.194).

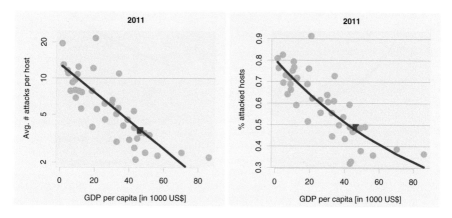

Fig. 5.187   Average number of attacks per host (*left*) and percentage of attacked host (*right*). *Blue line*: predicted values based on GDP-only model

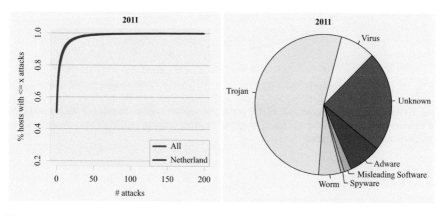

Figs. 5.188   Netherland: Empirical cumulative distribution of % of hosts with less than or equal to x attacks, and **5.189** Netherland: Distribution of attacks by type of malware

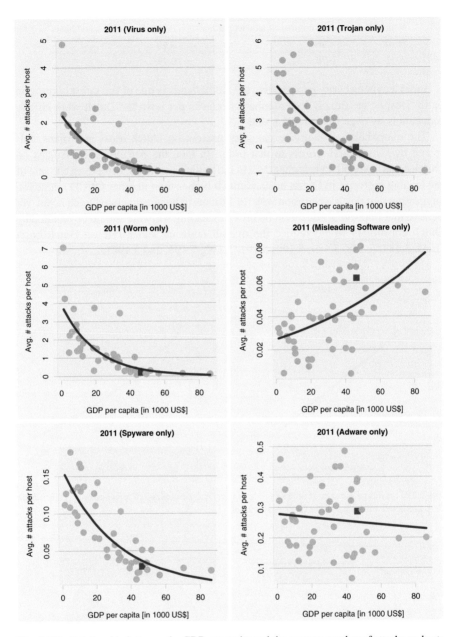

Fig. 5.190 Relationship between the GDP per capita and the average number of attacks on hosts of a country separately for virus, Trojan, worm, misleading software, spyware and adware attacks. Selected countries highlighted

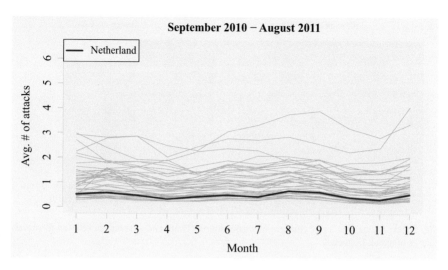

Fig. 5.191 Timeline of average monthly attacks per host by country for the 2011 time period

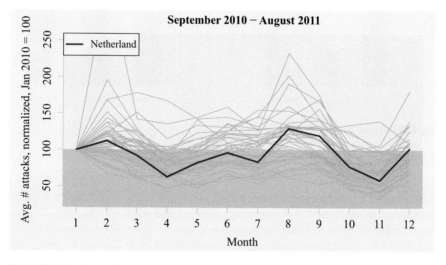

Fig. 5.192 Timeline of average monthly attacks per host by country for the 2011 time period. Normalized by attack count in September 2010

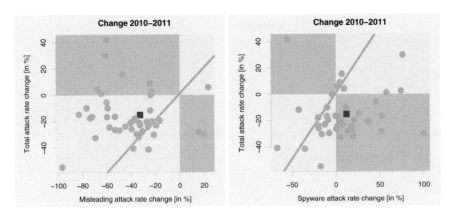

Fig. 5.193 Attack frequency changes of misleading software and spyware in relation to overall change of attack frequency

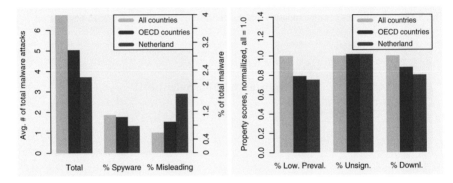

Fig. 5.194 Attack frequencies and host properties

5.25   New Zealand

New Zealand's cybersecurity strategy [34], published in 2011, provides a detailed description of the Kiwis' cybersecurity concerns and methods. Because 75% of New Zealanders have Internet access at home, over 70% have broadband at home, and over 77% use Internet banking, it is essential for New Zealand to protect its digital users. The report states that over 70% of users in New Zealand have been the victims of cybercrime and that over 59% have no security on their mobile phones.

The NZ cybersecurity strategy has three major goals: (1) increasing awareness of online security, (2) protecting government systems, and (3) responding to incidents. In order to achieve the first goal, the New Zealand government has partnered with an NGO called NetSafe to promote and educate the population about safe cybersecurity practices. In order to support the second goal, NZ has established a National Cyber Security Bureau within the Government Communications Security Bureau, the NZ analog of the US National Security Agency. In addition, in order to fulfil the third goal, the GCSB has established the NCSB and various CERTs.

New Zealand's vulnerability to malware is described via the following statistical tables.

New Zealand	Avg number of attacks per host	Percentage of attacked hosts
2010	5.07	0.53
2011	2.97	0.45

NZ has achieved an over 40% reduction in the number of attacks per host (on average) in just 1 year—by this metric, NZ was the seventh safest country in 2011. This is a remarkable achievement in just 1 year—but the fact that the number of Attacked Hosts was only reduced by about 15% suggests that it is possible to achieve even further reductions.

New Zealand is primarily targeted by Trojans, followed by worms and viruses. Moreover, New Zealanders are more heavily targeted by misleading software than others with a similar GDP—and additionally, the percentage of malware in NZ that is due to spyware and misleading software is higher than in similar countries, suggesting that an education campaign that focuses on the risks of misleading software and spyware could potentially reduce the malware rate on NZ hosts (Figs. 5.195, 5.196, 5.197, 5.198, 5.199, 5.200, 5.201, and 5.202).

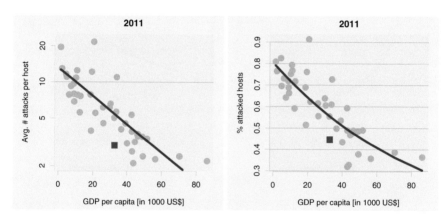

Fig. 5.195 Average number of attacks per host (*left*) and percentage of attacked host (*right*). *Blue line*: predicted values based on GDP-only model

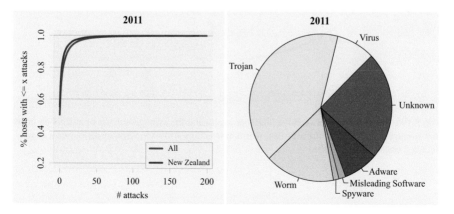

Figs. 5.196 New Zealand: Empirical cumulative distribution of % of hosts with less than or equal to x attacks, and **5.197** New Zealand: Distribution of attacks by type of malware

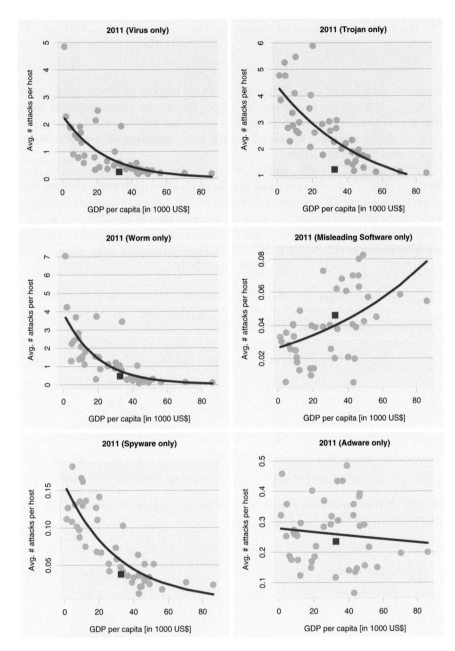

Fig. 5.198 Relationship between the GDP per capita and the average number of attacks on hosts of a country separately for virus, Trojan, worm, misleading software, spyware and adware attacks. Selected countries highlighted

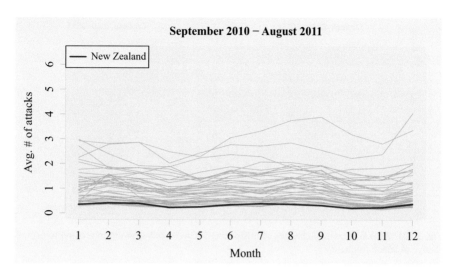

Fig. 5.199 Timeline of average monthly attacks per host by country for the 2011 time period

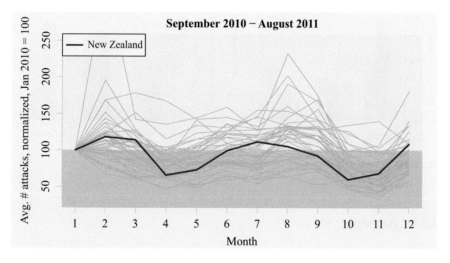

Fig. 5.200 Timeline of average monthly attacks per host by country for the 2011 time period. Normalized by attack count in September 2010

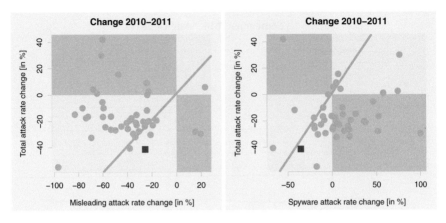

Fig. 5.201 Attack frequency changes of misleading software and spyware in relation to overall change of attack frequency

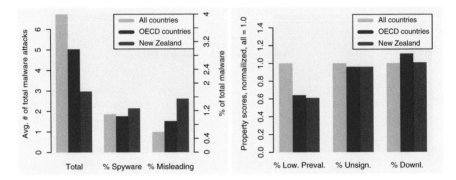

Fig. 5.202 Attack frequencies and host properties

5.26   Norway

Norway's cybersecurity strategy [35] was released in 2012 and was developed jointly by four ministries—defense, justice and public security, transport and communication, and government administration, reform, and church affairs. The work built on a previous information security strategy developed in 2003, as well as the work of multiple subsequent committees focused on identifying vulnerabilities and protecting infrastructure. It is not surprising that Norway has one of the best records in the cybersecurity space.

Norway's cybersecurity strategy has the four goals of ensuring: (1) awareness of cyber risk at both an individual and organizational level, (2) protection of national ICT infrastructure, (3) establishing security awareness and processes within public and private organizations, and (4) individuals are encouraged and trained to protect their personal data.

In order to achieve these goals, Norway has identified seven priority areas for further action. A comprehensive approach to information security involves developing a combination of technical, process, and regulatory instruments with the goal of ensuring that all organizations have a systematic approach to cybersecurity. Another priority area is to protect critical infrastructure such as the power grid and communication networks. Another priority is to ensure a uniform approach to security in government systems such as using electronic IDs, authentication and encryption techniques, and more. The development of instruments to detect cyber-criminal transactions (e.g. online malware marketplaces) and appropriate legislative authority to prosecute such actions is another priority. The government has also recognized the need for investment in cutting edge research and development in cybersecurity, and has established several such research programs.

With such a comprehensive and long-term policy, Norway has done well in protecting itself from cyber-attacks as shown in the table below.

Norway	Avg number of attacks per host	Percentage of attacked hosts
2010	2.87	0.44
2011	2.21	0.36

Simply put, Norway is one of the safest countries in the world from a cybersecurity point of view. The primary type of threat in Norway is in the form of Trojans, followed by viruses and worms. Interestingly, the percentage of pieces of spyware on Norwegian hosts is slightly higher than in other countries with a similar GDP. In addition, misleading software is more prevalent on Norwegian hosts than on other OECD countries. Thus, despite Norway's success in combating malware, there is still room for improvement along these two dimensions (Figs. 5.203, 5.204, 5.205, 5.206, 5.207, 5.208, 5.209, and 5.210).

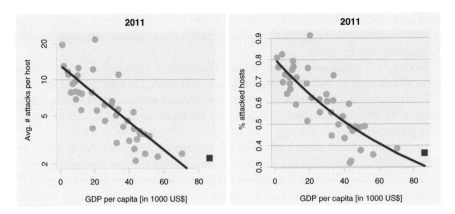

Fig. 5.203 Average number of attacks per host (*left*) and percentage of attacked host (*right*). *Blue line*: predicted values based on GDP-only model

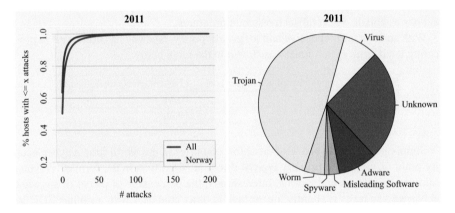

Figs. 5.204 Norway: Empirical cumulative distribution of % of hosts with less than or equal to x attacks, and **5.205** Norway: Distribution of attacks by type of malware

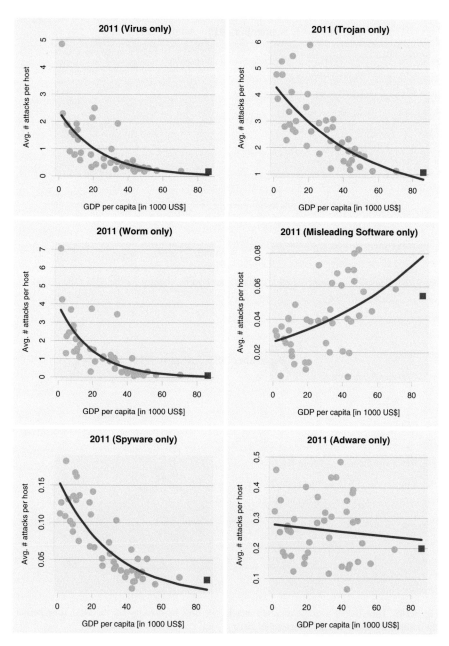

Fig. 5.206 Relationship between the GDP per capita and the average number of attacks on hosts of a country separately for virus, Trojan, worm, misleading software, spyware and adware attacks. Selected countries highlighted

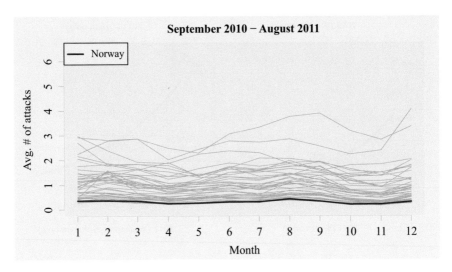

Fig. 5.207 Timeline of average monthly attacks per host by country for the 2011 time period

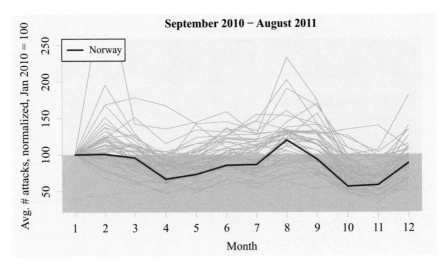

Fig. 5.208 Timeline of average monthly attacks per host by country for the 2011 time period. Normalized by attack count in September 2010

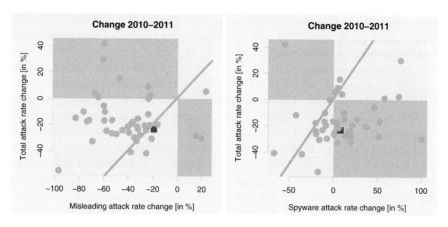

Fig. 5.209 Attack frequency changes of misleading software and spyware in relation to overall change of attack frequency

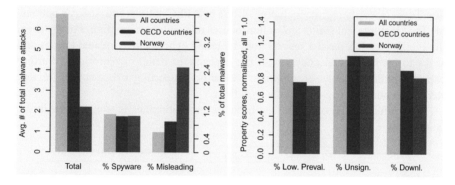

Fig. 5.210 Attack frequencies and host properties

5.27   Philippines

The Philippines' National Cyber Security Plan (NCSP) [36, 37] was first released in 2005 and is part of their National Critical Infrastructure Protection Plan (NCIPP). The plan divides up cyberspace in the Philippines into several parts: (1) enterprise networks, (2) Local Internet Service Providers, (3) Wide Area Network Providers, (4) the Internet Backbone consisting of entities such as phone companies, (5) entities that provide services to users such as content and telecom services, and (6) end users. The report identifies several high level threat vectors including cyber-crime, cyber-terrorism, foreign intelligence operations, techno-terrorism (in which a kinetic attack targets computational capabilities), and hacktivism. The report identifies several types of specific attack vectors that the Philippines needs to protect against. Overall, the goal of NCSP is to effectively protect critical infrastructure from cyber-attack, build appropriate capacity to predict and respond to attacks, develop appropriate legislative and law enforcement tools, and help educate the population about the importance of cybersecurity.

In order to address these goals, the NCSP has four pillars: (1) the "Understanding the Risk" component which plans to "map" out the cyber-geography of the Philippines, develop methods to monitor various types of traffic and online phenomena, assess risk. (2) the "Control the Risk" component includes some truly novel steps including the creation of a Cyber Special Operations Unit that carries out intelligence gathering, the production of national intelligence estimates covering cyber threats on a monthly basis, the development of a database of information about hackers, and developing appropriate educational curricula. In addition, this step involves the creation of multiple CERTs and establishment of a National Computer Forensics Lab, and establishing appropriate recovery programs. (3) The "Organization and Mobilization for Cybersecurity" pillar establishes the Task Force for the Security of Critical Infrastructure (TFSCI) as the focal point for cybersecurity of the country, together with a variety of regional centers, and operations. The Philippines NCSP expresses support for the importance of being part of an international cybersecurity regime including efforts by APEC nations and ASEAN nations. (4) The "Institutional Build-Up" pillar focuses on institution building. For instance, the Philippines has passed several laws targeting cybersecurity-related crimes and seeks to create special courts with the capacity to handle cybercrimes. Further capacity building looks at adopting standards, supporting cybersecurity related R&D, and building capacity in the workforce.

The overall profile of attacks on hosts in the Philippines is summarized in the table below.

Philippines	Avg number of attacks per host	Percentage of attacked hosts
2010	10	0.73
2011	12.99	0.76

The Philippines is one of the few countries that experienced a sharp uptick in the average number of Attacked Hosts (almost 30%) from 2010 to 2011—however, during the same period, the percentage of Filipino hosts rose by a much smaller amount, suggesting that Filipino hosts were attacked at increasingly high rates in 2011 as compared to 2010.

Unlike most countries, the principal attack vector in the Philippines consists of worms, followed by Trojans, and then viruses. The rate of adware in the Philippines is much higher than other countries with a similar GDP. Moreover, in terms of cyber-hygiene of the population, Filipino users tend to have much higher fractions of downloaded binaries (and a somewhat higher fraction of low-prevalence binaries) than hosts in other countries. Simply put, this suggests that educating Filipinos about the risks of low-prevalence binaries and downloaded binaries, as well as the ability to identify adware, could help mitigate the cyber-threat to the Philippines (Figs. 5.211, 5.212, 5.213, 5.214, 5.215, 5.216, 5.217, and 5.218).

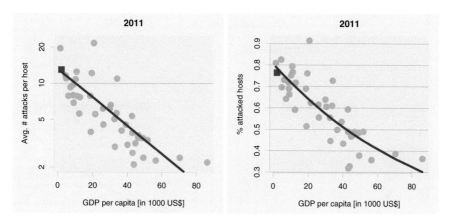

Fig. 5.211 Average number of attacks per host (*left*) and percentage of attacked host (*right*). *Blue line*: predicted values based on GDP-only model

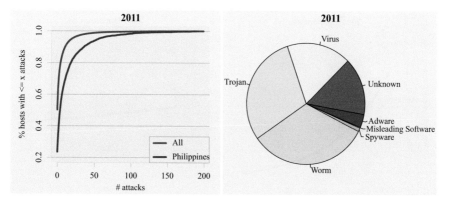

Figs. 5.212 Philippines: Empirical cumulative distribution of % of hosts with less than or equal to x attacks, and **5.213** Philippines: Distribution of attacks by type of malware

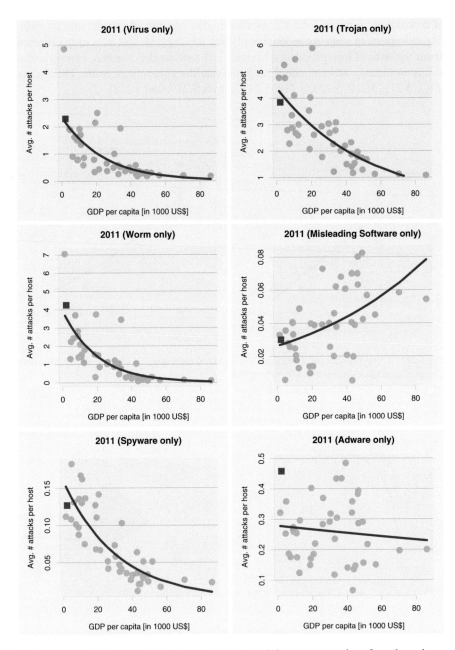

Fig. 5.214 Relationship between the GDP per capita and the average number of attacks on hosts of a country separately for virus, Trojan, worm, misleading software, spyware and adware attacks. Selected countries highlighted

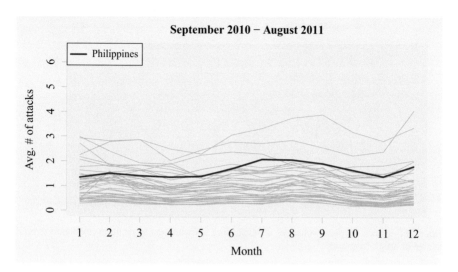

Fig. 5.215 Timeline of average monthly attacks per host by country for the 2011 time period

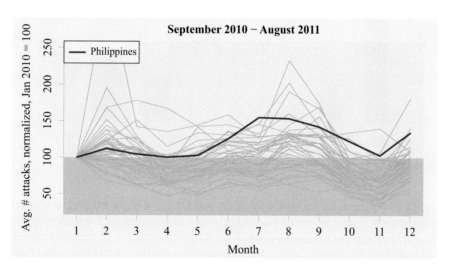

Fig. 5.216 Timeline of average monthly attacks per host by country for the 2011 time period. Normalized by attack count in September 2010

5.28 Poland

Poland's Cybersecurity Protection Plan (CPP) [38] was created by the Ministry of Administration and Digitization in partnership with Poland's Internal Security Agency (ABW for short) which is Poland's main internal intelligence agency, the counterpart of the FBI in the USA.

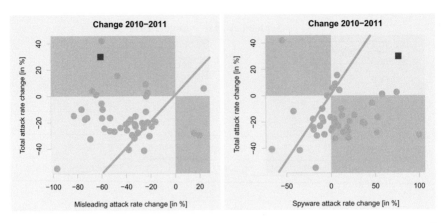

Fig. 5.217 Attack frequency changes of misleading software and spyware in relation to overall change of attack frequency

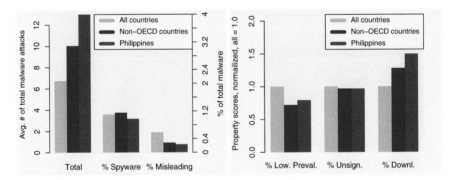

Fig. 5.218 Attack frequencies and host properties

The CPP focuses on unclassified systems and has the goals of reducing attacks on government ICT systems, reduce the number of cyber-attacks, improve capacity in cybersecurity, and improve the management of the cybersecurity threat. Ensuring the correct and continuous functioning of Polish government infrastructure is a key element of the CPP. The Government Cyber Security Incident Response Team (cert.gov.pl) is the principal CERT in Poland—a separate organization handles this function for the military.

The Polish Government requires that each government ministry submit a report analyzing cyber-risk within the country in January of each year, thus forcing an annual threat analysis exercise across the government. In addition to establishing a legal and law enforcement framework within Poland, the Government also has identified plenipotentiaries for cyberspace who have the responsibility of assessing risk, assessing methods to handle threats, conducting periodic training exercises, and ensuring various types of compliance. Education efforts envisaged in the CPP include cybersecurity training as part of higher education curricula as well as programs directed at children, youth, parents, teachers and government employees. Moreover, the CPP recognizes that multiple stakeholders need to be involved in cybersecurity.

The table below summarizes the cybersecurity situation in Poland.

Poland	Avg number of attacks per host	Percentage of attacked hosts
2010	6.77	0.65
2011	5.59	0.59

Poland did well in reducing the average number of attacks per host by over 17% and achieving an almost 10% reduction in the percentage of attacked hosts.

The primary nature of the malware threat to Poland consists of Trojans, followed in that order, by worms and viruses. The percentage of Polish hosts attacked by spyware is larger than the percentage for OECD countries. This suggests that educating Polish users about spyware—how it is delivered and how to recognize it—could help reduce such attacks in the country (Figs. 5.219, 5.220, 5.221, 5.222, 5.223, 5.224, 5.225, and 5.226).

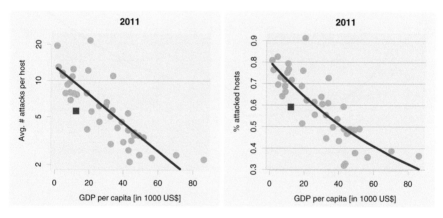

Fig. 5.219 Average number of attacks per host (*left*) and percentage of attacked host (*right*). *Blue line*: predicted values based on GDP-only model

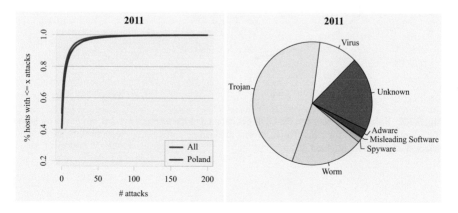

Figs. 5.220 Poland: Empirical cumulative distribution of % of hosts with less than or equal to x attacks, and **5.221** Poland: Distribution of attacks by type of malware

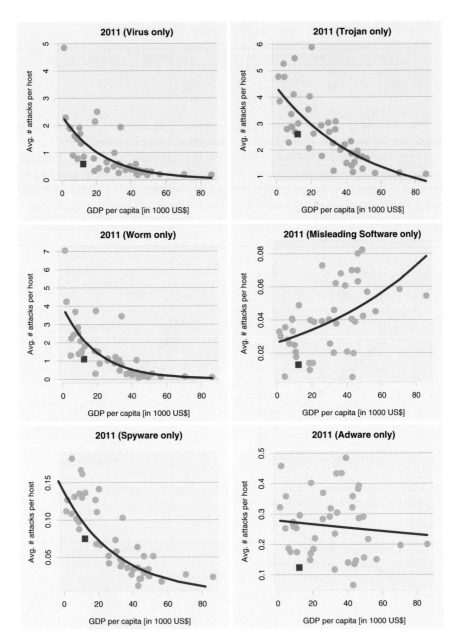

Fig. 5.222 Relationship between the GDP per capita and the average number of attacks on hosts of a country separately for virus, Trojan, worm, misleading software, spyware and adware attacks. Selected countries highlighted

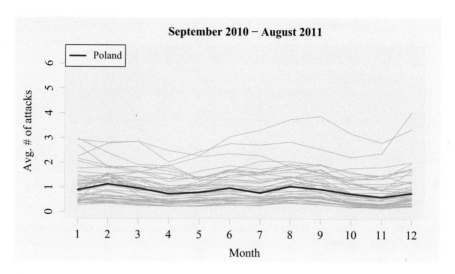

Fig. 5.223 Timeline of average monthly attacks per host by country for the 2011 time period

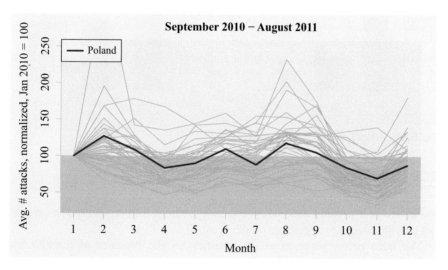

Fig. 5.224 Timeline of average monthly attacks per host by country for the 2011 time period. Normalized by attack count in September 2010

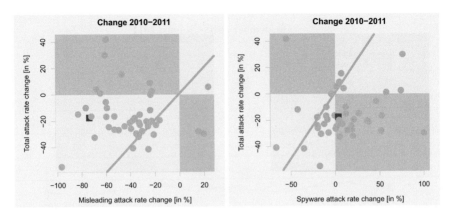

Fig. 5.225 Attack frequency changes of misleading software and spyware in relation to overall change of attack frequency

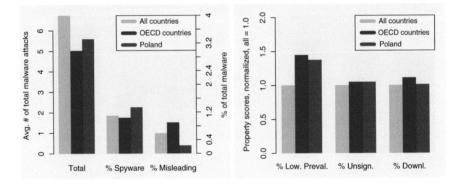

Fig. 5.226 Attack frequencies and host properties

5.29 Portugal

We were unable to find a national cybersecurity strategy for Portugal and hence cannot provide concrete suggestions on the strategy.

The table below summarizes the statistics for the presence of malware on Portuguese hosts.

Portugal	Avg number of attacks per host	Percentage of attacked hosts
2010	7.04	0.68
2011	5.54	0.62

We see that from 2010 to 2011, there was a reduction of over 20% in the average number of attacks per host—during the same time, there was a reduction of almost 10% of the percentage of attacked hosts. This suggests that—for whatever reason—Portuguese hosts are becoming more resilient to attack.

As is common for many countries, the principal types of malware targeting Portuguese hosts are Trojans, worms, and viruses, in that order. Portuguese hosts are more susceptible to misleading software than hosts in countries with a similar GDP. We note that the percentage of spyware on Portuguese hosts is greater than that in other OECD countries. One interesting point to note is that Portuguese hosts seem to have a lot of low-prevalence binaries. The data therefore suggests that Portuguese cybersecurity can be enhanced by educational efforts aimed at recognizing spyware and misleading software, as well as at pointing out the risks of downloading low prevalence binaries (Figs. 5.227, 5.228, 5.229, 5.230, 5.231, 5.232, 5.233, and 5.234).

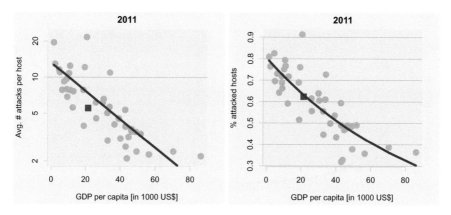

Fig. 5.227 Average number of attacks per host (*left*) and percentage of attacked host (*right*). *Blue line*: predicted values based on GDP-only model

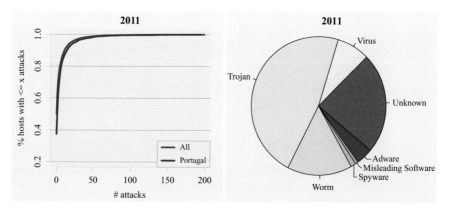

Figs. 5.228 Portugal: Empirical cumulative distribution of % of hosts with less than or equal to x attacks, and **5.229** Portugal: Distribution of attacks by type of malware

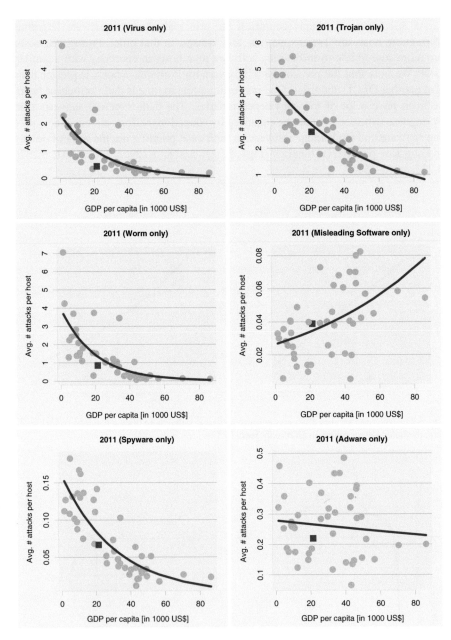

Fig. 5.230 Relationship between the GDP per capita and the average number of attacks on hosts of a country separately for virus, Trojan, worm, misleading software, spyware and adware attacks. Selected countries highlighted

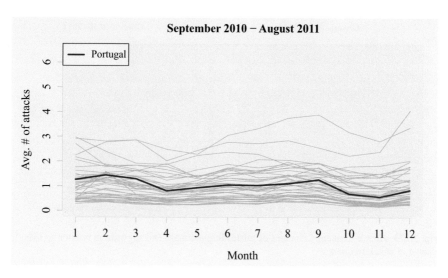

Fig. 5.231 Timeline of average monthly attacks per host by country for the 2011 time period

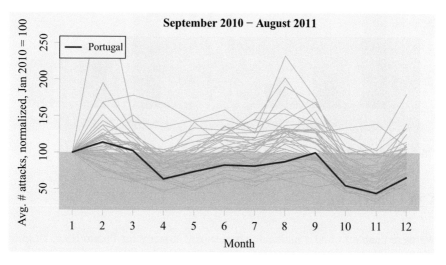

Fig. 5.232 Timeline of average monthly attacks per host by country for the 2011 time period. Normalized by attack count in September 2010

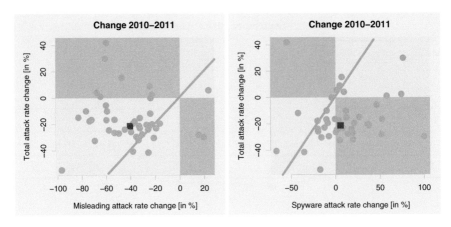

Fig. 5.233 Attack frequency changes of misleading software and spyware in relation to overall change of attack frequency

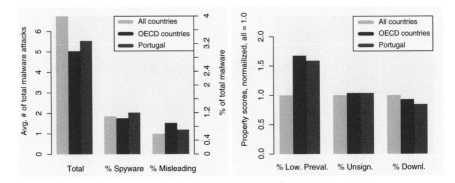

Fig. 5.234 Attack frequencies and host properties

5.30 Puerto Rico

We were unable to find a national cybersecurity strategy for Puerto Rico. Despite this, the table below shows that Puerto Rican hosts did well in terms of cybersecurity compared to other countries.

Puerto Rico	Avg number of attacks per host	Percentage of attacked hosts
2010	5.68	0.60
2011	4.55	0.56

We see immediately that Puerto Rico was subject to fewer attacks per host than other countries with a similar GDP, and moreover, the percentage of attacked hosts in Puerto Rico was also smaller than their GDP might have led one to believe.

Puerto Rican hosts are primarily targeted by Trojans, worms, and viruses, in that order. We note that Puerto Rico was much more heavily targeted by misleading software and adware than hosts in countries with a similar GDP. In addition, the percentage of Puerto Rican hosts with misleading software was higher than in OECD countries. The data therefore suggests that cybersecurity of Puerto Rican hosts can be improved by education efforts that show users how to recognize and/or remove misleading software and adware (Figs. 5.235, 5.236, 5.237, 5.238, 5.239, 5.240, 5.241, and 5.242).

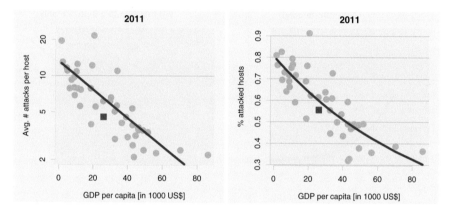

Fig. 5.235 Average number of attacks per host (*left*) and percentage of attacked host (*right*). *Blue line*: predicted values based on GDP-only model

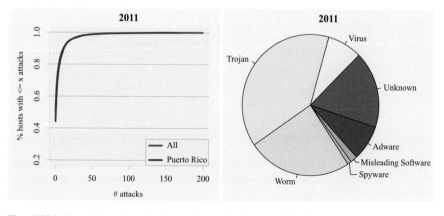

Figs. 5.236 Puerto Rico: Empirical cumulative distribution of % of hosts with less than or equal to x attacks, and **5.237** Puerto Rico: Distribution of attacks by type of malware

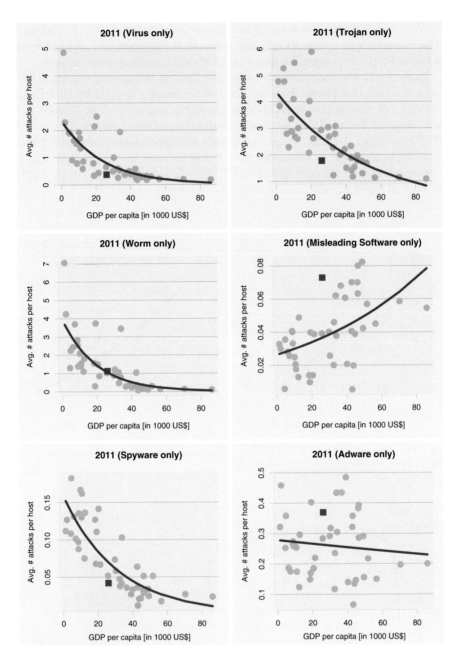

Fig. 5.238 Relationship between the GDP per capita and the average number of attacks on hosts of a country separately for virus, Trojan, worm, misleading software, spyware and adware attacks. Selected countries highlighted

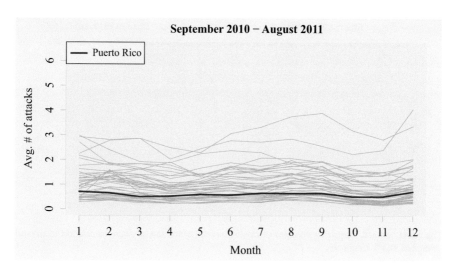

Fig. 5.239 Timeline of average monthly attacks per host by country for the 2011 time period

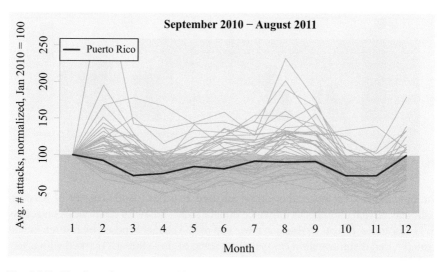

Fig. 5.240 Timeline of average monthly attacks per host by country for the 2011 time period. Normalized by attack count in September 2010

5.31 Russia

Russia's Information Security Doctrine [39] was published in late December 2008, though President Putin's approval date is listed as being in September 2000. The document lists multiple information security related threats to Russia including threats to individuals and groups, threats to policies of the Russian state, threats to Russia's IT industry, and threats to Russian facilities both within and outside Russia. The document lists several types of threats of concern which include monopolies of

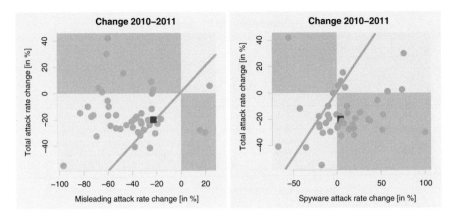

Fig. 5.241 Attack frequency changes of misleading software and spyware in relation to overall change of attack frequency

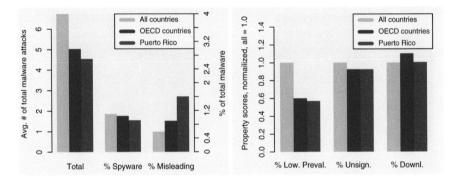

Fig. 5.242 Attack frequencies and host properties

various industry sectors, cyber-crime targeted at privacy and security of Russian nationals, illegal attempts to influence populations*, unlawful restrictions on access to IT resources, threats to Russian news and media channels posed by foreign media outlets*, and disinformation campaigns*. Some of these threats, marked with asterisks in the previous sentence, appear in the Russian Information Security Doctrine, but in no other national cybersecurity document that we have seen, denoting the different perception of the cyber-threat that is seen by Russian policymakers.

In order to address these risks, the Russian cybersecurity policy proposes a mix of legal, technological, and economic measures. The economic measures, in particular, relate to the availability of financing to support the creation of law-enforcement infrastructure and cybersecurity related technological advances.

The table below summarizes the statistics about cyber-attacks on hosts within Russia.

Russia	Avg number of attacks per host	Percentage of attacked hosts
2010	12.50	0.81
2011	12.50	0.79

As the above table clearly shows, Russia is highly vulnerable to cyber-attacks according to both measures listed above and in both years. About 80% of Russian hosts are attacked and on average, Russian hosts have 12.5 pieces of malware on them.

Russian hosts are primarily attacked by Trojans, viruses, and worms, in that order. Though Russia's GDP is not very high, Russian hosts are more vulnerable than nations with similar GDPs. Russian hosts tend to have more low-prevalence binaries and more downloaded binaries than hosts from other non-OECD countries. All of these suggest that education campaigns that seek to emphasize how to recognize Trojans, and the risk of low prevalence and downloaded binaries could help reduce the number of cyber-attacks in Russia. World-leading Russian cybersecurity firms like Kaspersky Lab could play an important role in shaping such education and outreach efforts (Figs. 5.243, 5.244, 5.245, 5.246, 5.247, 5.248, 5.249, and 5.250).

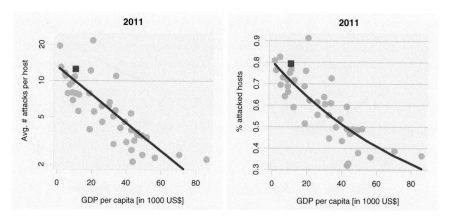

Fig. 5.243 Average number of attacks per host (*left*) and percentage of attacked host (*right*). *Blue line*: predicted values based on GDP-only model

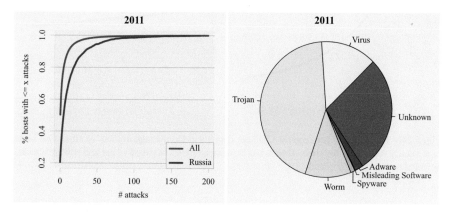

Figs. 5.244 Russia: Empirical cumulative distribution of % of hosts with less than or equal to x attacks, and **5.245** Russia: Distribution of attacks by type of malware

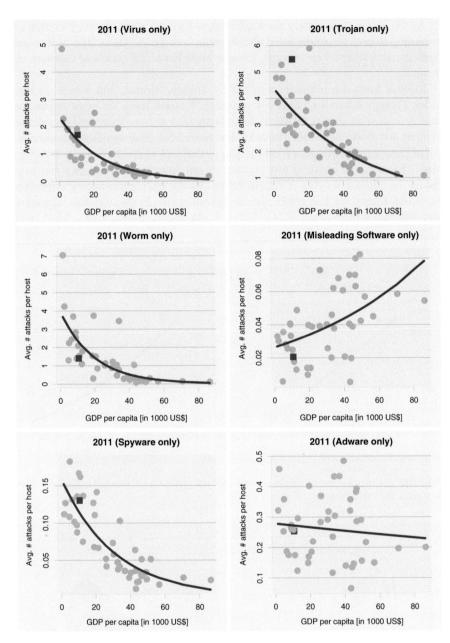

Fig. 5.246 Relationship between the GDP per capita and the average number of attacks on hosts of a country separately for virus, Trojan, worm, misleading software, spyware and adware attacks. Selected countries highlighted

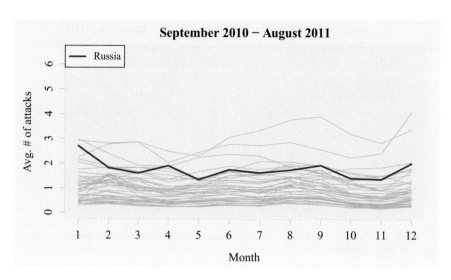

Fig. 5.247 Timeline of average monthly attacks per host by country for the 2011 time period

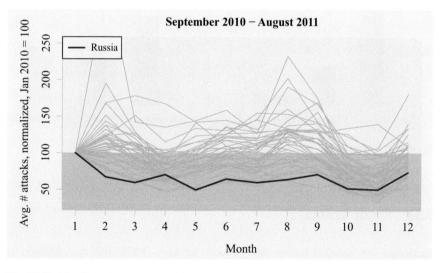

Fig. 5.248 Timeline of average monthly attacks per host by country for the 2011 time period. Normalized by attack count in September 2010

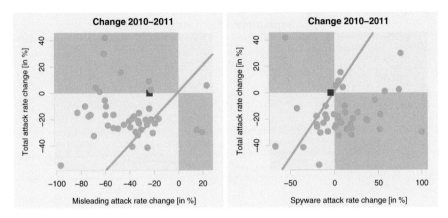

Fig. 5.249 Attack frequency changes of misleading software and spyware in relation to overall change of attack frequency

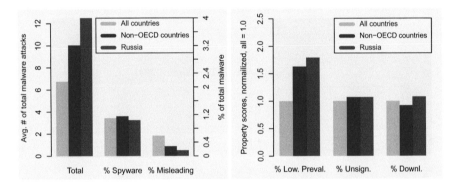

Fig. 5.250 Attack frequencies and host properties

5.32   Saudi Arabia

Saudi Arabia's National Information Security Strategy (NISS) [40] was released in 2011 and focuses on (1) supporting the free flow of information, (2) ensuring the security of information, (3) developing robust and resilient ICT systems, (4) raising awareness and building capacity around cyber issues and (5) creating appropriate national guidelines. In order to achieve these goals, the Saudi government has developed a 10-point plan that will: (1) build an effective and secure information environment, (2) ICT infrastructure hardening and enhancement, (3) building human capacity in cyber-security, (4) developing capabilities to collect vulnerabilities and resources to mitigate threat, (5) develop methods to mitigate risk, (6) develop compliance guidelines and methods to track compliance related metrics, (7) supporting research and entrepreneur-ship around cybersecurity, (8) develop processes to identify threat and help support Saudi Arabia's defensive posture, (9) better coordination with international partners, and (10) raise awareness of best cyber-hygiene practices amongst the population.

Critical infrastructure protection is a key aspect of Saudi cybersecurity—for instance, the mega oil-company Saudi Aramco was the target of the Shamoon mal-ware in 2012 which launched a DDOS attack that destroyed over 30,000 disk drives [41]. The NISS specifically points out the lack of protection against cyber attacks on SCADA systems within Saudi Arabia—critical for the country's economy.

The table below summarizes the statistics we have on Saudi Arabian hosts.

Saudi Arabia	Avg number of attacks per host	Percentage of attacked hosts
2010	14.77	0.81
2011	12.17	0.76

Though Saudi Arabia is one of the most vulnerable nations on earth from a cybersecurity perspective, there was a sharp drop in the average number of attacks per host from 2010 to 2011, as well as a significant drop in the percentage of attacked hosts from 2010 to 2011. Nonetheless, the vulnerability of Saudi Arabia to cyber-attacks is greater than that of other nations with a similar GDP. In fact, this held for every type of malware that we studied, suggesting that the country can do better.

The main attack vector on Saudi hosts are Trojans, followed by worms and viruses in that order. Cyber-attacks through low-prevalence binaries and down-loaded binaries seem higher in Saudi Arabia than in other non-OECD countries, suggesting that education strategies aimed at educating consumers on how to detect these two malware distribution methods could help alleviate some of Saudi Arabia's cyber-risk (Figs. 5.251, 5.252, 5.253, 5.254, 5.255, 5.256, 5.257, and 5.258).

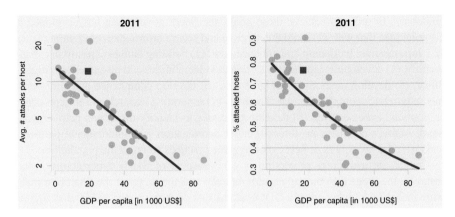

Fig. 5.251 Average number of attacks per host (*left*) and percentage of attacked host (*right*). *Blue line*: predicted values based on GDP-only model

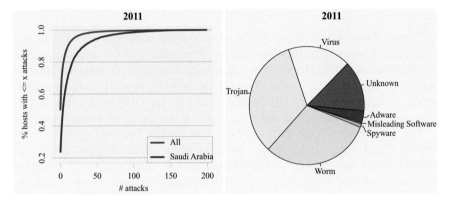

Figs. 5.252 Saudi Arabia: Empirical cumulative distribution of % of hosts with less than or equal to x attacks, and **5.253** Saudi Arabia: Distribution of attacks by type of malware

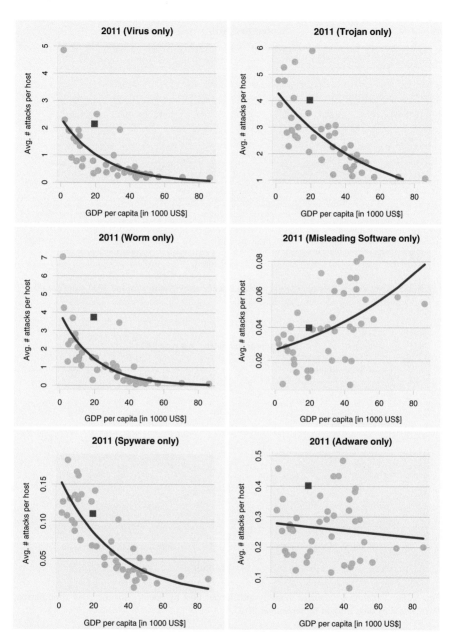

Fig. 5.254 Relationship between the GDP per capita and the average number of attacks on hosts of a country separately for virus, Trojan, worm, misleading software, spyware and adware attacks. Selected countries highlighted

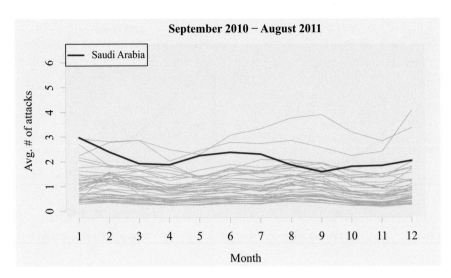

Fig. 5.255 Timeline of average monthly attacks per host by country for the 2011 time period

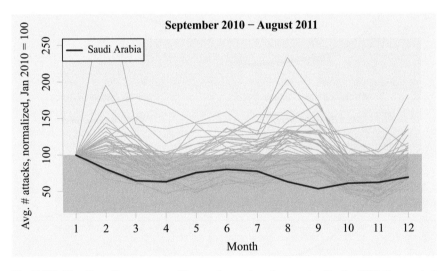

Fig. 5.256 Timeline of average monthly attacks per host by country for the 2011 time period.
Normalized by attack count in September 2010

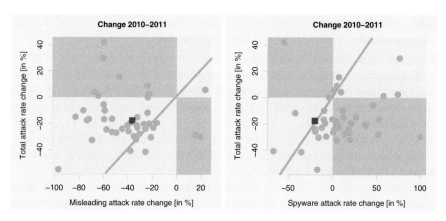

Fig. 5.257 Attack frequency changes of misleading software and spyware in relation to overall change of attack frequency

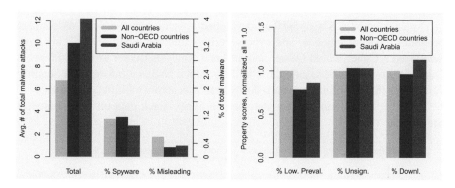

Fig. 5.258 Attack frequencies and host properties

5.33 Singapore

Singapore's Information Security Masterplan two was launched in 2008 with the goal of boosting investor confidence in Singapore as a cyber-safe hub for international trade and commerce. A more recent Masterplan called the National Cyber Security Masterplan 2018 [42, 43] identifies social engineering, spear-phishing, and advanced persistent threats as major challenges. The report identifies "supply chain hacking" in which small/medium enterprises are hacked in order to obtain compromising data about their (larger) customers. The Masterplan cites specific attacks as examples of the risk to Singapore.

Masterplan 2018 identifies (1) securing public communications systems infrastructure, systems and services (2) education and awareness of cyber-risk amongst consumers, (3) awareness and guidance in helping businesses better protect their enterprises, and (4) hardening the critical infrastructure in Singapore, as critical goals.

In order to achieve these goals, Singapore plans to increase domestic and international collaborations in cybersecurity, enhance human capital through improved education, make investments in cybersecurity related R&D, and take steps that catalyze the development and implementation of advanced cybersecurity practices in industry.

Organizationally, Singapore has developed specific cybersecurity assessment programs, code of practice, and cybersecurity exercises. Additionally, they have developed a Cyber Watch Center and a Threat Analysis Center to track threats, a DigiSafe program for education and training, and a National Information Security Committee that coordinates cybersecurity practices across multiple ministries.

The table below summarizes statistics on cyber-attack data on hosts based in Singapore.

Singapore	Avg number of attacks per host	Percentage of attacked hosts
2010	6.79	0.65
2011	5.36	0.59

Though Singapore is one of the most advanced nations in the world from a cybersecurity perspective, we see that its attack rate, according to both measures above, is higher than that in nations with a comparable GDP.

Trojans, worms, and viruses, in that order, pose the biggest malware vectors in Singapore. Spyware and adware are also higher in Singaporean hosts than the nation's GDP would suggest. Misleading software seems more prevalent than expected, as is the number of low-prevalence binaries and downloaded binaries on Singaporean hosts. All of these suggest that Singapore's excellent cyber-education programs have a few specific targets to focus on which might further help mitigate cyber-risk in the country (Figs. 5.259, 5.260, 5.261, 5.262, 5.263, 5.264, 5.265, and 5.266).

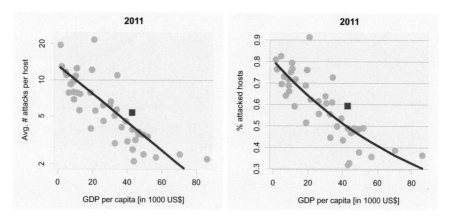

Fig. 5.259 Average number of attacks per host (*left*) and percentage of attacked host (*right*). *Blue line*: predicted values based on GDP-only model

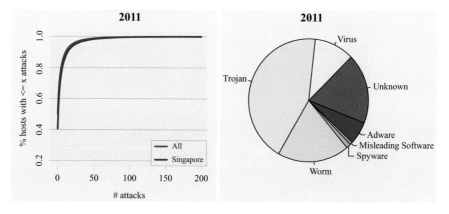

Figs. 5.260 Singapore: Empirical cumulative distribution of % of hosts with less than or equal to x attacks, and **5.261** Singapore: Distribution of attacks by type of malware

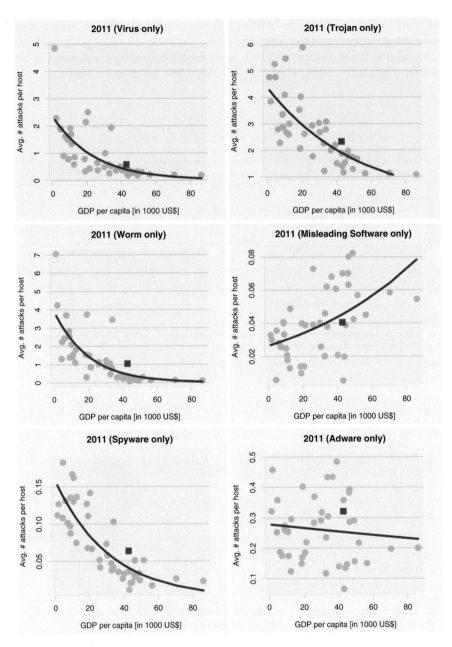

Fig. 5.262 Relationship between the GDP per capita and the average number of attacks on hosts of a country separately for virus, Trojan, worm, misleading software, spyware and adware attacks. Selected countries highlighted

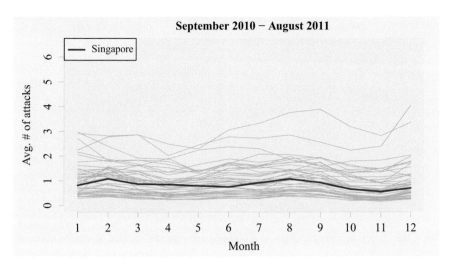

Fig. 5.263 Timeline of average monthly attacks per host by country for the 2011 time period

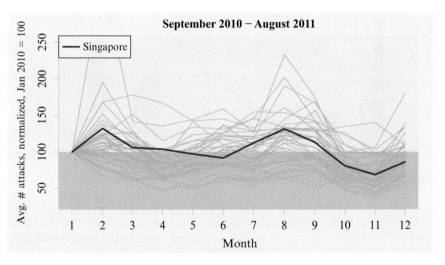

Fig. 5.264 Timeline of average monthly attacks per host by country for the 2011 time period. Normalized by attack count in September 2010

5.34 South Africa

Though we were unable to get a finalized copy of South Africa's cybersecurity policy, we were able to get a draft policy dated February 2010 [44]. Interestingly, this document is the only national cybersecurity policy that we saw that explicitly states that the country involved lags behind other nations in securing cyberspace. Despite this observation, our data shows that South Africa is far from being the worst in terms of cyber-vulnerability.

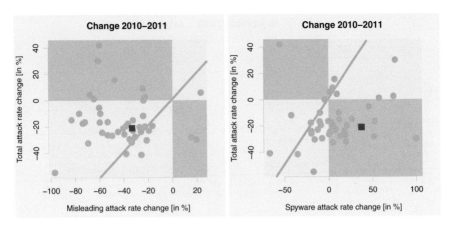

Fig. 5.265 Attack frequency changes of misleading software and spyware in relation to overall change of attack frequency

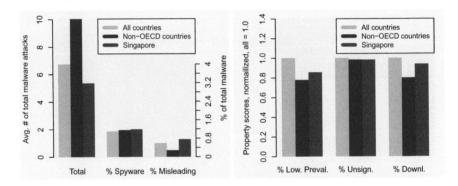

Fig. 5.266 Attack frequencies and host properties

The goals of South Africa's cybersecurity policy are to help build relevant structures in support of cybersecurity, reduce threats and vulnerabilities, build both public private partnerships and international collaborations, develop capacity, and build understand/comply with international technical standards.

In order to achieve these goals, South Africa proposes the creation of a National Cybersecurity Advisory Council that will coordinate cybersecurity activities around the country and take direct actions to support the goals referenced in the preceding paragraph. In addition, the government proposes creating a National Computer Security Incident Response Team (CSIRT), a government CSIRT, as well as specific teams devoted to various sectors. The South African draft policy recognizes the need to facilitate both partnerships between the state and private entities, as well as international cooperation, and specifies forums and groups that the CSIRTs above will engage with. Capacity building and skills development is also an important area of focus.

The basic statistics about South Africa's vulnerability to cyber-attack are described in the table below.

South Africa	Avg number of attacks per host	Percentage of attacked hosts
2010	6.79	0.65
2011	5.36	0.59

The table shows that South Africa did quite well in terms of cybersecurity, achieving a significant reduction from 2010 to 2011 in terms of both the average number of attacks per host and the percentage of attacked hosts.

The principal cyber-threat to South Africa consists of worms, followed by Trojans and viruses. Except for worms, South Africa is less vulnerable to all other forms of cyber-attack than one would expect from a country with its GDP. Behaviorally, South African hosts tend to have a larger number of low-prevalence binaries and a significantly larger number of downloaded binaries than non-OECD countries on average, suggesting that an education effort focused on explaining the risks of low-prevalence and downloaded binaries would help improve South Africa's cyber defenses (Figs. 5.267, 5.268, 5.269, 5.270, 5.271, 5.272, 5.273, and 5.274).

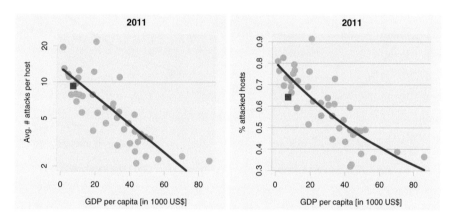

Fig. 5.267 Average number of attacks per host (*left*) and percentage of attacked host (*right*). *Blue line*: predicted values based on GDP-only model

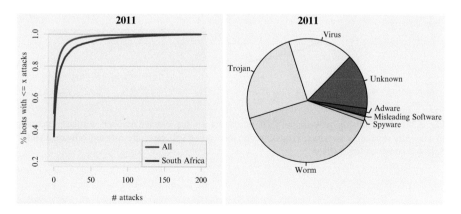

Figs. 5.268 South Africa: Empirical cumulative distribution of % of hosts with less than or equal to x attacks, and **5.269** South Africa: Distribution of attacks by type of malware

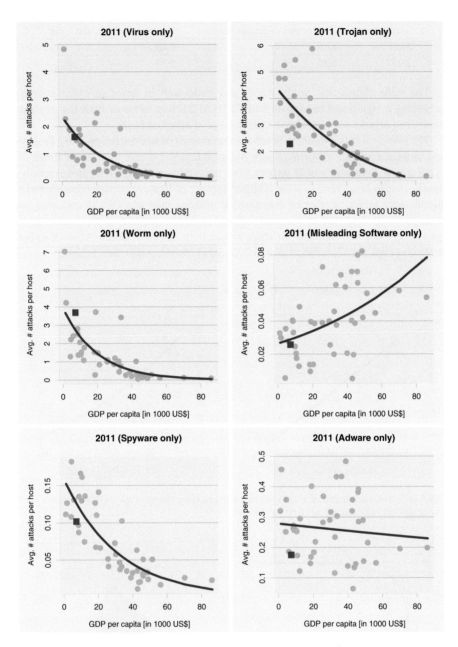

Fig. 5.270 Relationship between the GDP per capita and the average number of attacks on hosts of a country separately for virus, Trojan, worm, misleading software, spyware and adware attacks. Selected countries highlighted

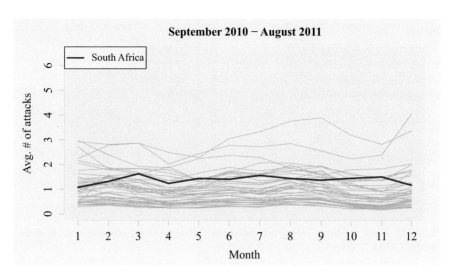

Fig. 5.271 Timeline of average monthly attacks per host by country for the 2011 time period

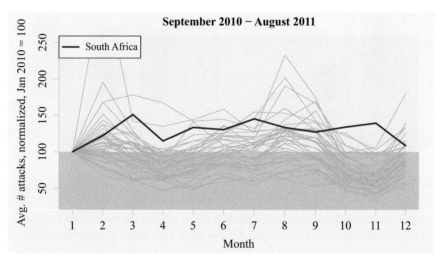

Fig. 5.272 Timeline of average monthly attacks per host by country for the 2011 time period. Normalized by attack count in September 2010

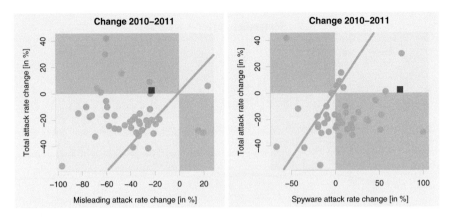

Fig. 5.273 Attack frequency changes of misleading software and spyware in relation to overall change of attack frequency

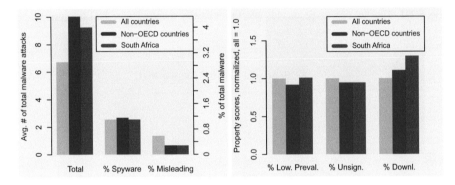

Fig. 5.274 Attack frequencies and host properties

5.35   South Korea

South Korea competes with India for the distinction of being the most cyber-vulnerable country in the world. In 91% of South Korean hosts were attacked by malware and on average had 21.63 attacks—a staggering rate for a major developed economy.

South Korea's brief cybersecurity Masterplan [45] was released in August 2011. Created by the National Cyber Security Strategy Council, the plan envisages a joint military, public, and private response framework, protecting critical infrastructure, national level detection and nullification of attacks, achieving deterrence through international partnerships, and building a cybersecurity infrastructure.

The document suggests setting up a 3-ringed defense at the international gateway level, the internet service provider level, and the end-user level. It also emphasizes a specific focus on cybersecurity for financial organizations, and envisages the distribution of anti-virus software to the private sector.

The table below summarizes the state of cyber-attacks against South Korean hosts.

South Korea	Avg number of attacks per host	Percentage of attacked hosts
2010	15.24	0.82
2011	21.63	0.91

South Korea is primarily targeted by Trojans, followed by viruses and worms, though, interestingly, it is also the subject of many types of attacks by unknown types of malware. It has been suggested to the author that many of these attacks may be explicit attacks by a hostile North Korean regime—though we do not have any data to either substantiate or repudiate such allegations. Specifically, spyware, Trojans, and viruses are found on South Korean hosts in far larger abundance than one would expect from a nation with South Korea's GDP (Figs. 5.275, 5.276, 5.277, 5.278, 5.279, 5.280, 5.281, and 5.282).

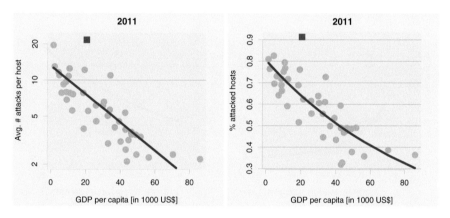

Fig. 5.275 Average number of attacks per host (*left*) and percentage of attacked host (*right*). *Blue line*: predicted values based on GDP-only model

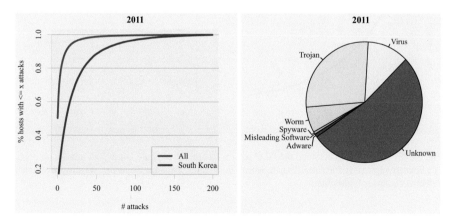

Figs. 5.276 South Korea: Empirical cumulative distribution of % of hosts with less than or equal to x attacks, and **5.277** South Korea: Distribution of attacks by type of malware

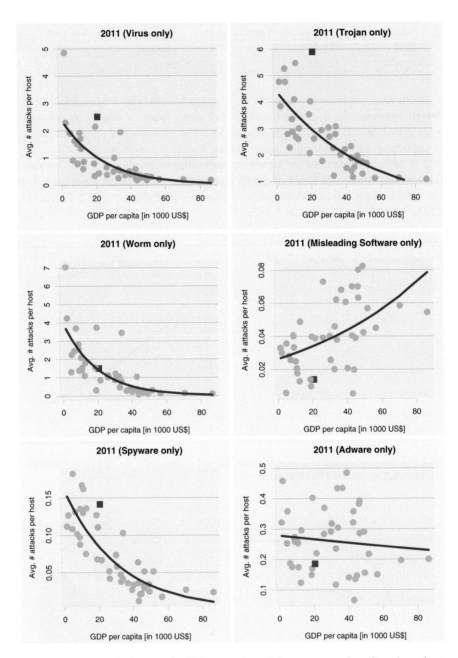

Fig. 5.278 Relationship between the GDP per capita and the average number of attacks on hosts of a country separately for virus, Trojan, worm, misleading software, spyware and adware attacks. Selected countries highlighted

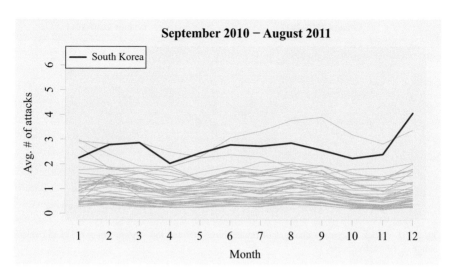

Fig. 5.279 Timeline of average monthly attacks per host by country for the 2011 time period

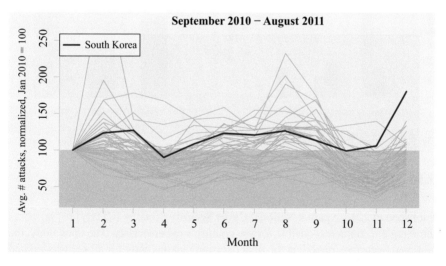

Fig. 5.280 Timeline of average monthly attacks per host by country for the 2011 time period. Normalized by attack count in September 2010

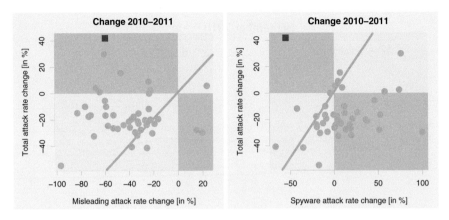

Fig. 5.281 Attack frequency changes of misleading software and spyware in relation to overall change of attack frequency

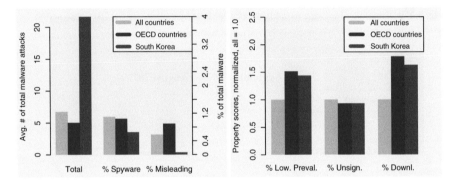

Fig. 5.282 Attack frequencies and host properties

5.36   Spain

Published in 2013, Spain's National Cyber Security Strategy (CITE) [46] proposes three organizations responsible for the country's cybersecurity. The apex body, the National Security Council, is the final decision making body for all security related issues including cybersecurity. The Specialized Cyber Security Committee is responsible for coordinating cybersecurity matters between different state organizations and entities as well as the private sector. The Specialized Situation Committee is charged with handling emerging cybersecurity events.

Spain's cybersecurity strategy has several elements. Foremost amongst them is to ensure that the nation's communications infrastructure is safe for both government institutions and private institutions, and is resilient to attack. Another major element is to develop the capacity, technology, and institutions needed for law enforcement to protect Spain from both cyber-crime and terrorist attacks. Raising public awareness of cybersecurity best practices, and building capacity in this domain across the board are key elements, as is the development of international partnerships. Spain lays additional emphasis of collaboration with EU cybersecurity entities.

The table below summarizes the vulnerability of Spanish hosts to cyber-attacks.

Spain	Avg number of attacks per host	Percentage of attacked hosts
2010	8.36	0.74
2011	6.15	0.64

Our data shows a significant improvement from 2010 to 2011 in Spain's vulnerability to cyber attacks. The average number of attacks per host dropped by over 26%, while the percentage of attacked hosts dropped by over 13%—significant improvements over a very short time period.

The principal vectors used to carry out cyber-attacks on Spain are Trojans, worms, and viruses (in that order). Spain is slightly more vulnerable to Trojans, adware, and spyware than other countries with comparable GDPs (Figs. 5.283, 5.284, 5.285, 5.286, 5.287, 5.288, 5.289, and 5.290).

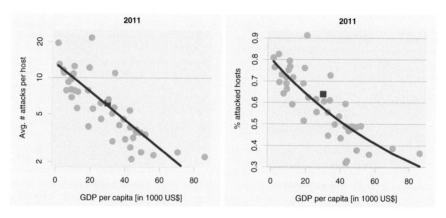

Fig. 5.283 Average number of attacks per host (*left*) and percentage of attacked host (*right*). *Blue line*: predicted values based on GDP-only model

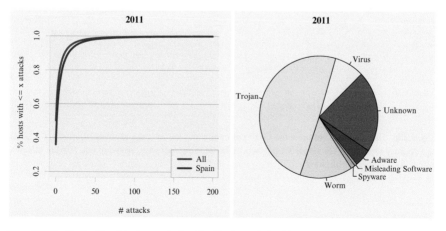

Figs. 5.284 Spain: Empirical cumulative distribution of % of hosts with less than or equal to x attacks, and **5.285** Spain: Distribution of attacks by type of malware

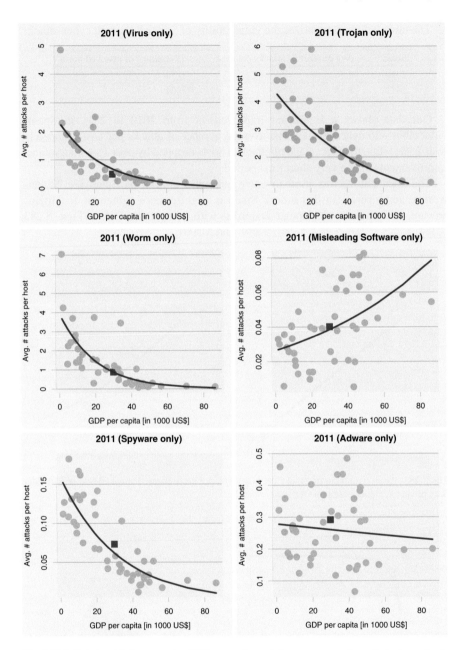

Fig. 5.286 Relationship between the GDP per capita and the average number of attacks on hosts of a country separately for virus, Trojan, worm, misleading software, spyware and adware attacks. Selected countries highlighted

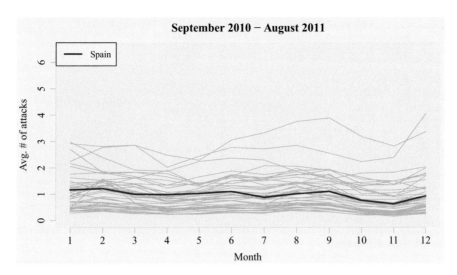

Fig. 5.287 Timeline of average monthly attacks per host by country for the 2011 time period

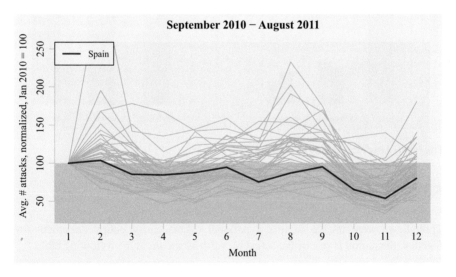

Fig. 5.288 Timeline of average monthly attacks per host by country for the 2011 time period. Normalized by attack count in September 2010

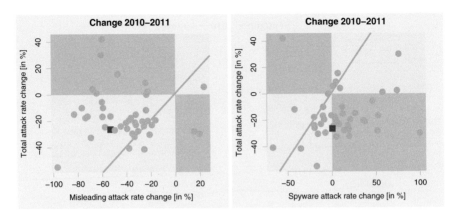

Fig. 5.289 Attack frequency changes of misleading software and spyware in relation to overall change of attack frequency

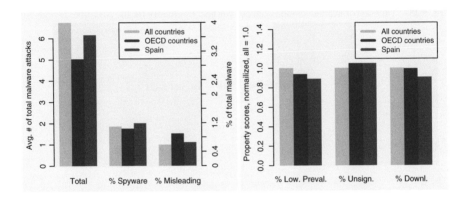

Fig. 5.290 Attack frequencies and host properties

5.37 Sweden

Though we were unable to find a national cybersecurity policy document for Sweden, the nation is one of the safest on earth as shown in the following table.

Sweden	Avg number of attacks per host	Percentage of attacked hosts
2010	3.17	0.46
2011	2.41	0.38

The main modalities of attacks on Swedish hosts are Trojans, followed by viruses and worms, in that order. Swedish hosts are safer than other countries with a similar GDP with respect to all types of malware that we studied. The one place for improvement is in the case of misleading software where the percentage of attacked Swedish hosts exceeds that of other OECD countries. Nonetheless, Sweden is one of the safest countries in the world from a cybersecurity perspective (Figs. 5.291, 5.292, 5.293, 5.294, 5.295, 5.296, 5.297, and 5.298).

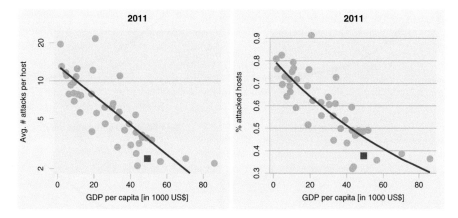

Fig. 5.291 Average number of attacks per host (*left*) and percentage of attacked host (*right*). *Blue line*: predicted values based on GDP-only model

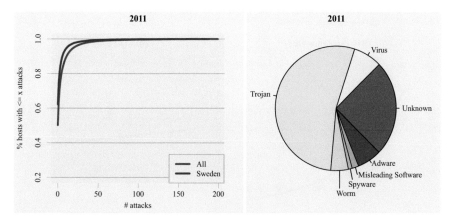

Figs. 5.292 Sweden: Empirical cumulative distribution of % of hosts with less than or equal to x attacks, and **5.293** Sweden: Distribution of attacks by type of malware

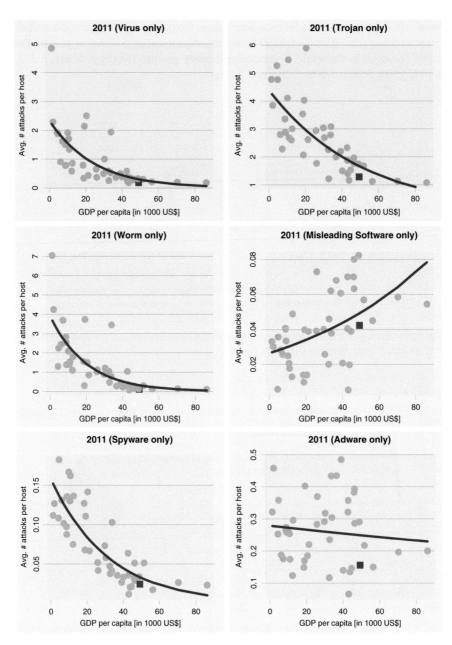

Fig. 5.294 Relationship between the GDP per capita and the average number of attacks on hosts of a country separately for virus, Trojan, worm, misleading software, spyware and adware attacks. Selected countries highlighted

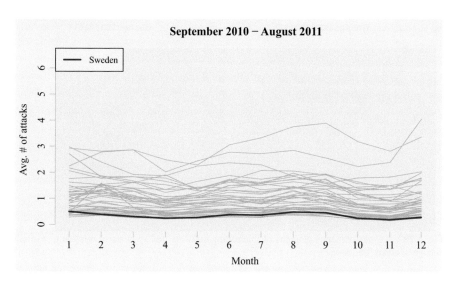

Fig. 5.295 Timeline of average monthly attacks per host by country for the 2011 time period

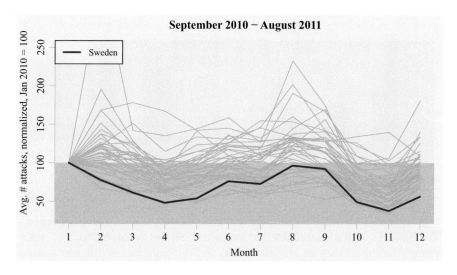

Fig. 5.296 Timeline of average monthly attacks per host by country for the 2011 time period. Normalized by attack count in September 2010

5.38   Switzerland

Switzerland's June 2012 cybersecurity strategy document [47] has three major goals: an early warning framework for cybersecurity threats, a strategy to make critical infrastructure systems more robust and resilient to cyber attacks, and the development of methods to reduce all types of malicious cyber activity including cyber-crime.

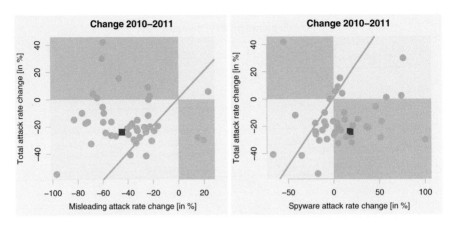

Fig. 5.297 Attack frequency changes of misleading software and spyware in relation to overall change of attack frequency

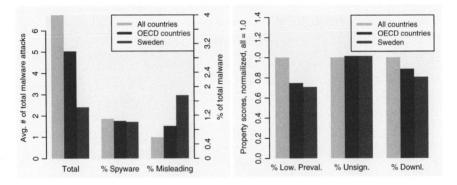

Fig. 5.298 Attack frequencies and host properties

In order to achieve these goals, Switzerland proposes seven "spheres of action". The first focuses on advanced R&D to anticipate different threats before they occur. A second sphere focuses on analysis of risks and vulnerabilities of different systems including critical infrastructure systems. A third focuses on analyzing past incidents and threats and developing the inter-cantonal mechanisms needed to coordinate responses between Switzerland's different cantons. A fourth focuses on developing capacity in the field. A fifth sphere focuses on ensuring Switzerland's participation in relevant international bodies and in developing the processes and protocols required to manage crises. Finally, the seventh sphere focuses on ensuring that Switzerland has the legal framework required to prosecute crimes in the cyber-domain.

Like Sweden, Switzerland is one of the safest nations on earth from a cybersecurity perspective. This is clear from the table below which summarizes cyber-attacks on Switzerland.

Switzerland	Avg number of attacks per host	Percentage of attacked hosts
2010	3.35	0.43
2011	2.41	0.39

Of all the different types of malware that threaten Switzerland, the most common ones are Trojans, viruses, and worms, in that order. The average number of attacks due to spyware is higher in Switzerland than one would expect from a nation with Switzerland's GDP. In addition, the percentage of Swiss hosts that are attacked by misleading software is higher than in the case of OECD countries (on average). This suggests that there is room for improvement through targeted cyber-education programs that educate Swiss citizens about how to recognize these two types of malware (Figs. 5.299, 5.300, 5.301, 5.302, 5.303, 5.304, 5.305, and 5.306).

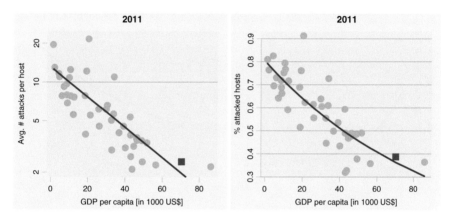

Fig. 5.299 Average number of attacks per host (*left*) and percentage of attacked host (*right*). *Blue line*: predicted values based on GDP-only model

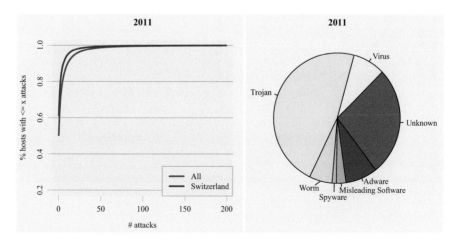

Figs. 5.300 Switzerland: Empirical cumulative distribution of % of hosts with less than or equal to x attacks, and **5.301** Switzerland: Distribution of attacks by type of malware

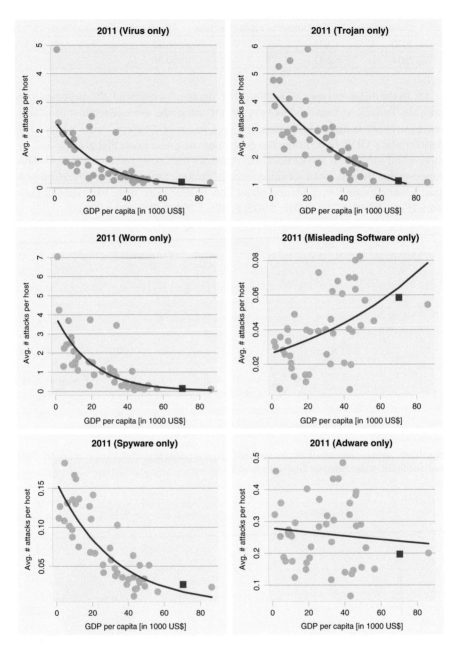

Fig. 5.302 Relationship between the GDP per capita and the average number of attacks on hosts of a country separately for virus, Trojan, worm, misleading software, spyware and adware attacks. Selected countries highlighted

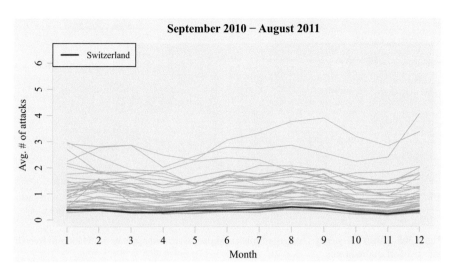

Fig. 5.303 Timeline of average monthly attacks per host by country for the 2011 time period

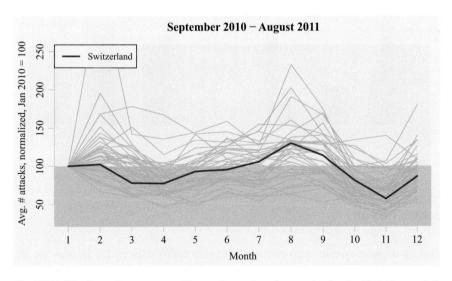

Fig. 5.304 Timeline of average monthly attacks per host by country for the 2011 time period.
Normalized by attack count in September 2010

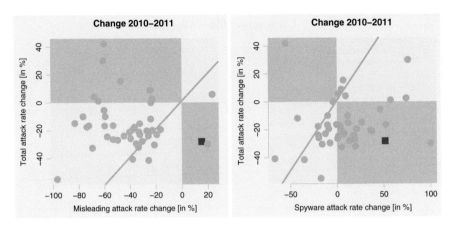

Fig. 5.305 Attack frequency changes of misleading software and spyware in relation to overall change of attack frequency

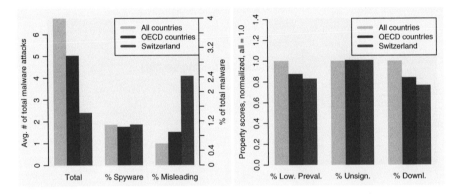

Fig. 5.306 Attack frequencies and host properties

5.39   Taiwan

Though we were unable to find a national cybersecurity strategy for Taiwan, we see from the table below that from 2010 to 2011, Taiwan was able to make significant improvements in its cyber-vulnerability.

Taiwan	Avg number of attacks per host	Percentage of attacked hosts
2010	9.40	0.75
2011	7.85	0.69

Trojans, worms, and viruses form the principal triple cyber-threat to Taiwan, in that order. The average number of attacks per host due to Trojans and spyware is higher in Taiwan than in other nations with a similar GDP. In addition, Taiwanese hosts tend to have more low-prevalence software and downloaded software than hosts in countries with a similar GDP. These facts suggest that Taiwan's cyber-education programs should place additional emphasis on how users may be infected by Trojans and by spyware, and focusing on good cyber-hygiene behaviors that shape behaviors linked to reducing the users' tendencies to download binaries and to download/install low-prevalence binaries (Figs. 5.307, 5.308, 5.309, 5.310, 5.311, 5.312, 5.313, and 5.314).

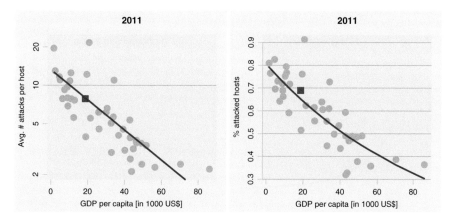

Fig. 5.307 Average number of attacks per host (*left*) and percentage of attacked host (*right*). *Blue line*: predicted values based on GDP-only model

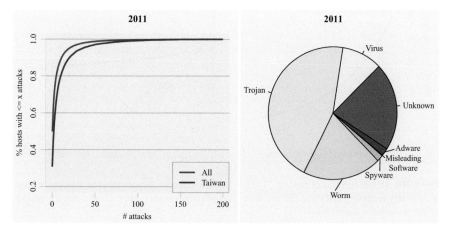

Figs. 5.308 Taiwan: Empirical cumulative distribution of % of hosts with less than or equal to x attacks, and **5.309** Taiwan: Distribution of attacks by type of malware

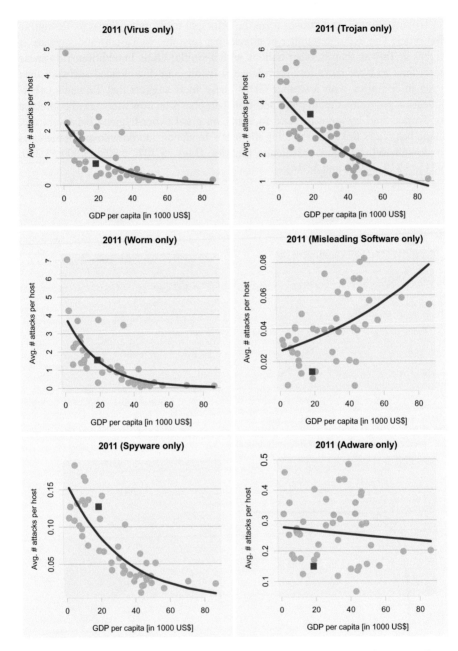

Fig. 5.310 Relationship between the GDP per capita and the average number of attacks on hosts of a country separately for virus, Trojan, worm, misleading software, spyware and adware attacks. Selected countries highlighted

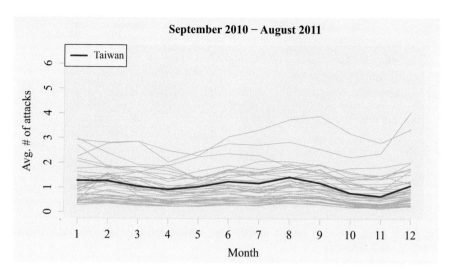

Fig. 5.311 Timeline of average monthly attacks per host by country for the 2011 time period

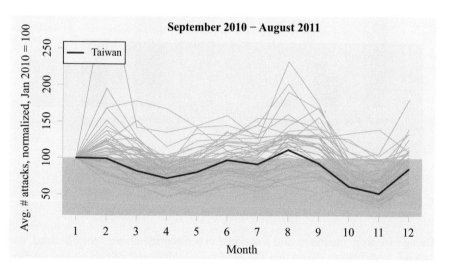

Fig. 5.312 Timeline of average monthly attacks per host by country for the 2011 time period. Normalized by attack count in September 2010

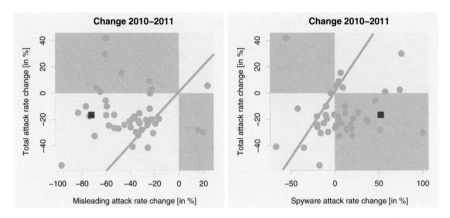

Fig. 5.313 Attack frequency changes of misleading software and spyware in relation to overall change of attack frequency

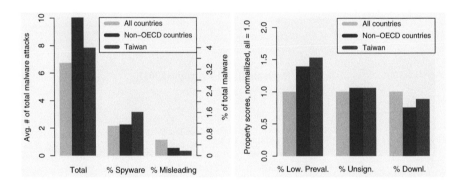

Fig. 5.314 Attack frequencies and host properties

5.40   Thailand

Though we were unable to find a national cybersecurity strategy for Thailand, our data (cf. table below) shows that Thailand has a high degree of cyber-vulnerability.

Thailand	Avg number of attacks per host	Percentage of attacked hosts
2010	10.57	0.70
2011	11.01	0.70

From 2010 to 2011, Thailand experienced a marginal increase in cyber-vulnerability, with Trojans, worms, and viruses forming the three major threats to Thailand. The average number of attacks on Thai hosts by Trojans, misleading software, adware, and spyware, all exceeded what we expect from a country with Thailand's GDP. Thai hosts also recorded a larger number of downloaded binaries and low-prevalence binaries than was true on average for OECD countries. These findings suggest that Thai cyber-education efforts should focus on recognizing Trojans, misleading software, adware, and spyware, and additionally teach users the risks of downloaded software and low-prevalence binaries (Figs. 5.315, 5.316, 5.317, 5.318, 5.319, 5.320, 5.321, and 5.322).

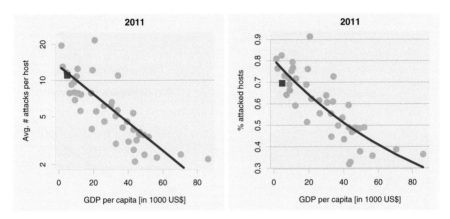

Fig. 5.315 Average number of attacks per host (*left*) and percentage of attacked host (*right*). *Blue line*: predicted values based on GDP-only model

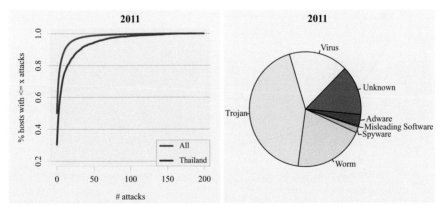

Figs. 5.316 Thailand: Empirical cumulative distribution of % of hosts with less than or equal to x attacks, and **5.317** Thailand: Distribution of attacks by type of malware

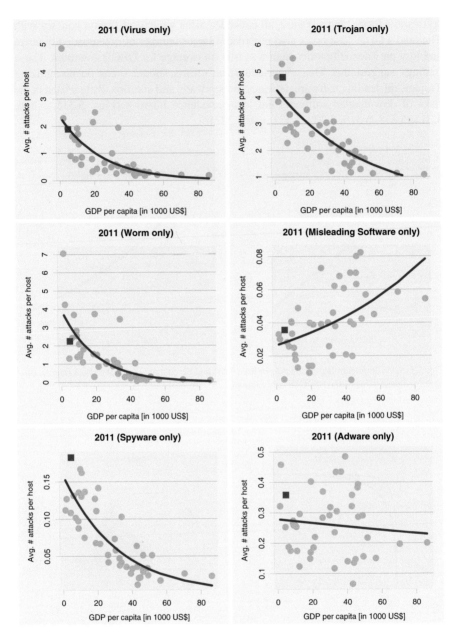

Fig. 5.318 Relationship between the GDP per capita and the average number of attacks on hosts of a country separately for virus, Trojan, worm, misleading software, spyware and adware attacks. Selected countries highlighted

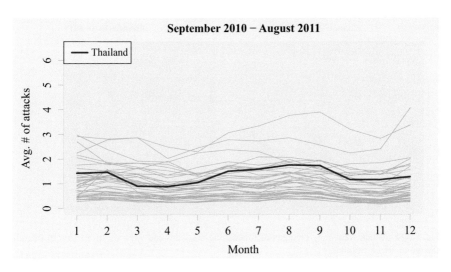

Fig. 5.319 Timeline of average monthly attacks per host by country for the 2011 time period

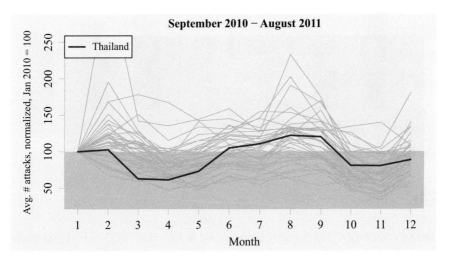

Fig. 5.320 Timeline of average monthly attacks per host by country for the 2011 time period. Normalized by attack count in September 2010

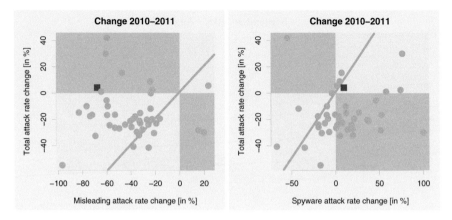

Fig. 5.321 Attack frequency changes of misleading software and spyware in relation to overall change of attack frequency

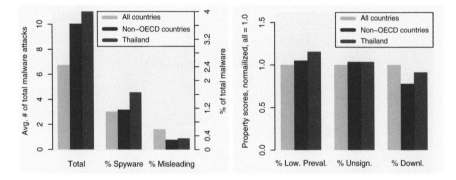

Fig. 5.322 Attack frequencies and host properties

5.41 Turkey

Turkey's National Cyber Security Strategy and Action Plan (2013–2014) [48] states that a Cyber Security Council has been established in the Ministry of Transport, Maritime Affairs, and Communication, with participation of senior officials of different ministries, intelligence organizations, and law enforcement agencies.

The goal of Turkey's cybersecurity policy is to (1) ensure the cybersecurity of data, processes and services provided by the government, (2) protect both public and private critical infrastructure, and (3) detect cyber incidents and recover from them as soon as possible, together with appropriate legal tools to bring those responsible to justice. An important aspect of Turkey's cybersecurity policy is the need for an integrated framework in which the technical dimension of cybersecurity is complemented by the economic, legal, administrative, political, and social dimensions. The need for cooperation between the public sector, private sector, academic sector, and international partners is also heavily emphasized.

The table below summarizes the data we have on attacks on Turkish hosts.

Turkey	Avg number of attacks per host	Percentage of attacked hosts
2010	10.70	0.78
2011	10.82	0.75

Trojans, worms, and viruses, in that order, form the three main threats to Turkish hosts. However, slightly differently, the average number of attacks per host in Turkey due to viruses, Trojans, and spyware, was greater in Turkish hosts than in other countries with a similar GDP (Figs. 5.323, 5.324, 5.325, 5.326, 5.327, 5.328, 5.329, and 5.330).

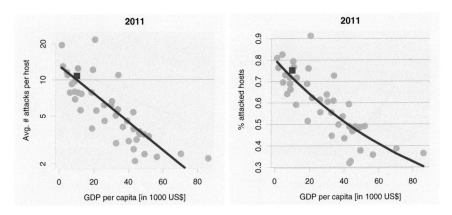

Fig. 5.323 Average number of attacks per host (*left*) and percentage of attacked host (*right*). *Blue line*: predicted values based on GDP-only model

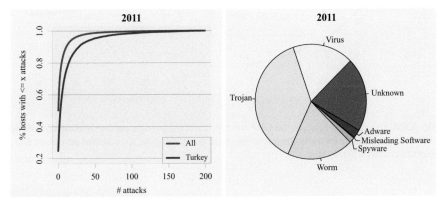

Figs. 5.324 Turkey: Empirical cumulative distribution of % of hosts with less than or equal to x attacks, and **5.325** Turkey: Distribution of attacks by type of malware

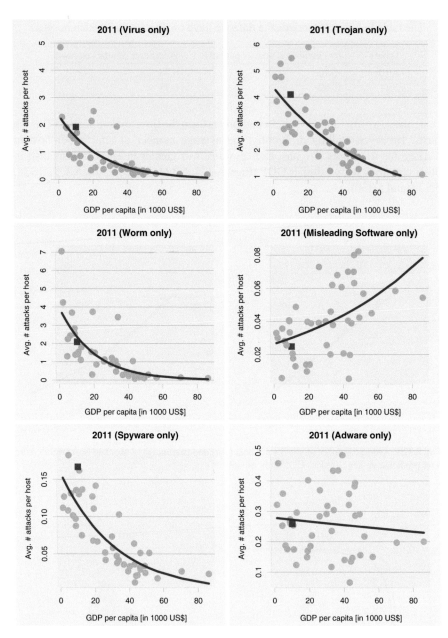

Fig. 5.326 Relationship between the GDP per capita and the average number of attacks on hosts of a country separately for virus, Trojan, worm, misleading software, spyware and adware attacks. Selected countries highlighted

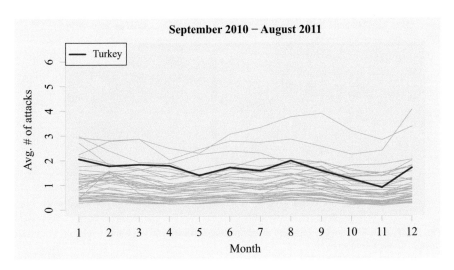

Fig. 5.327 Timeline of average monthly attacks per host by country for the 2011 time period

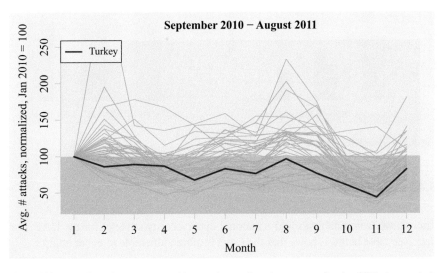

Fig. 5.328 Timeline of average monthly attacks per host by country for the 2011 time period. Normalized by attack count in September 2010

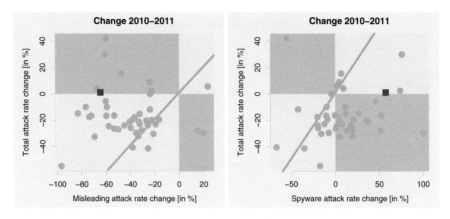

Fig. 5.329 Attack frequency changes of misleading software and spyware in relation to overall change of attack frequency

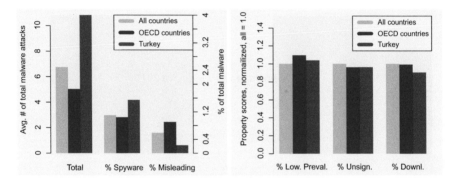

Fig. 5.330 Attack frequencies and host properties

5.42 United Arab Emirates

Though we were unable to find a national cybersecurity strategy for the UAE, our data (see table below) shows that the UAE is quite vulnerable to cyber attacks.

United Arab Emirates	Avg number of attacks per host	Percentage of attacked hosts
2010	9.51	0.70
2011	10.97	0.73

Unlike most countries in our study, the UAE's cyber-vulnerability increased from 2010 to 2011—in particular, there was an over 15% increase in the average number of attacks per host in just 1 year, even though the number of Attacked Hosts stayed the same. This suggests that hosts that were already attacked became even more vulnerable, even though 25–30% of hosts were unaffected.

Worms, Trojans, and viruses, in that order, were the three most significant cyber-threats to the UAE during this time. In all categories (except for misleading software), the UAE fell below the level of cybersecurity that one would expect for a country with their GDP, suggesting that the government needs to increase its emphasis on cybersecurity in the nation. Hosts in the UAE had a slightly higher number of low-prevalence binaries than hosts in OECD countries—and a significantly larger number of downloaded binaries than hosts in OECD countries, suggesting that cyber-education efforts could focus on highlighting the risks of these two behaviors (Figs. 5.331, 5.332, 5.333, 5.334, 5.335, 5.336, 5.337, and 5.338).

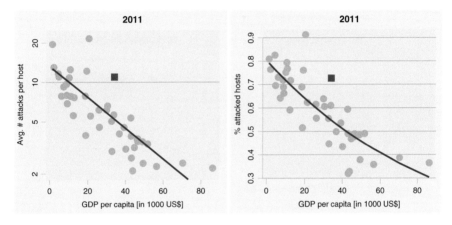

Fig. 5.331 Average number of attacks per host (*left*) and percentage of attacked host (*right*). *Blue line*: predicted values based on GDP-only model

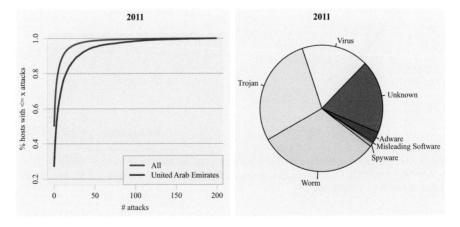

Figs. 5.332 United Arab Emirates: Empirical cumulative distribution of % of hosts with less than or equal to x attacks, and **5.333** United Arab Emirates: Distribution of attacks by type of malware

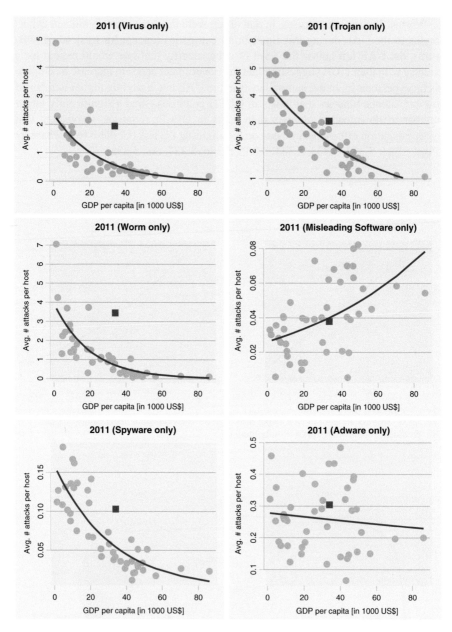

Fig. 5.334 Relationship between the GDP per capita and the average number of attacks on hosts of a country separately for virus, Trojan, worm, misleading software, spyware and adware attacks. Selected countries highlighted

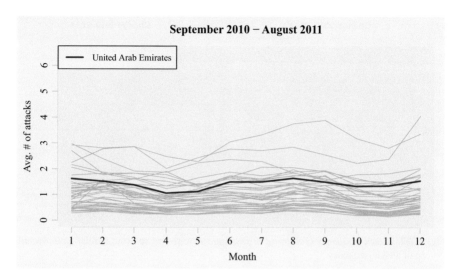

Fig. 5.335 Timeline of average monthly attacks per host by country for the 2011 time period

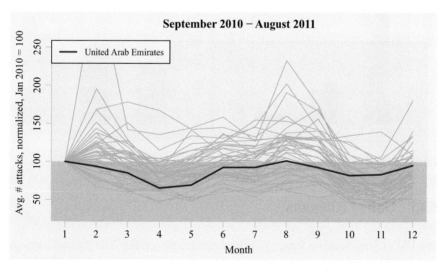

Fig. 5.336 Timeline of average monthly attacks per host by country for the 2011 time period. Normalized by attack count in September 2010

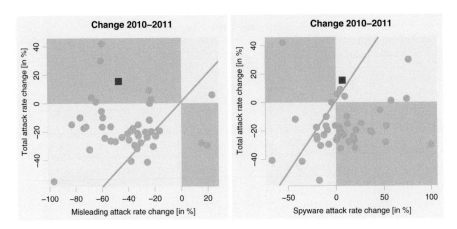

Fig. 5.337 Attack frequency changes of misleading software and spyware in relation to overall change of attack frequency

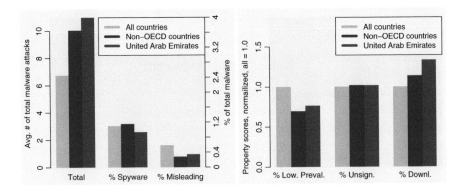

Fig. 5.338 Attack frequencies and host properties

5.43 United Kingdom

The UK's Cyber Security Strategy [49], released in November 2011, creates a vision that the UK hoped to achieve by 2015 which is when this report is going to the press. The main goal is to help the UK derive "huge economic and social value" [49, p. 7] by making the UK a safe cyber-haven. The report presents four goals: (1) to achieve sufficiently good cybersecurity and to counter cyber-crime sufficiently well that IT-enabled businesses can go about their missions safely, (2) to be more resilient and robust in the presence of cyber attacks, (3) to shape the UK's cyberspace in such a way that it is safe to use and so that it promotes open societies, and (4) to develop the skills and technology needed to achieve these goals. According to [49], over 20,000 malicious emails per month targeted UK government computers, making it an urgent national priority to reduce cyber attacks. To achieve these goals, the UK created a 4-year £650 (about $1B) National Cyber Security Program (NCSP) to achieve these goals—a yearly average of about $250 M. This program invests relatively small sums (under approximately $20 M each over the 4 year period) on building a cabinet office that maintains view of the operational threat, a service that builds secure government infrastructure, tackling cyber-crime, making cybersecurity an integral part of the Ministry of Defence, and working with the private sector to improve resilience. The bulk of the funds (59%) go to an intelligence account that maintains a cross-cutting cybersecurity capability headquartered at the UK's Government Communications Headquarters (GCHQ).

Specific objectives include (1) helping the GCHQ intelligence agency and Ministry of Defence better detect and neutralize threats to the UK, (2) better protect critical infrastructure including privately run infrastructure—including a Center for the Protection of National Infrastructure, (3) develop public-private partnerships that help secure private cyber-space and share threat information, (4) encourage development and adoption of industry-led cybersecurity standards, (5) help small businesses through the provision of advice and warnings, (6) strengthen legal frameworks and law enforcement frameworks, and more. One of the most interesting proposals is to help build a stronger partnership between the UK's key eavesdropping agency, the GCHQ, and businesses. Much like Israel's elite unit 8200 that has developed outstanding cybersecurity systems and expertise, this appears to be an effort by the UK to capitalize on cyber-attack and cyber-defense expertise within its GCHQ intelligence agency.

The table below summarizes our data about the cyber-threat to the UK.

United Kingdom	Avg number of attacks per host	Percentage of attacked hosts
2010	5.03	0.54
2011	4.05	0.50

Trojans dominate the threat landscape to the UK followed by viruses and worms, respectively. Compared to other countries with a similar GDP, the UK appears much more vulnerable to adware and misleading software. This suggests that a cyber-education program in the UK may focus on the risk of being attacked by fake disk cleanup and anti-virus utilities, as well as adware (Figs. 5.339, 5.340, 5.341, 5.342, 5.343, 5.344, 5.345, and 5.346).

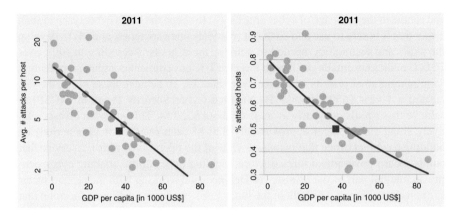

Fig. 5.339 Average number of attacks per host (*left*) and percentage of attacked host (*right*). *Blue line*: predicted values based on GDP-only model

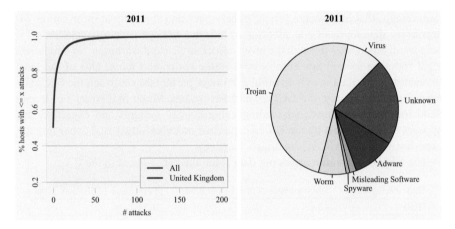

Figs. 5.340 United Kingdom: Empirical cumulative distribution of % of hosts with less than or equal to x attacks, and **5.341** United Kingdom: Distribution of attacks by type of malware

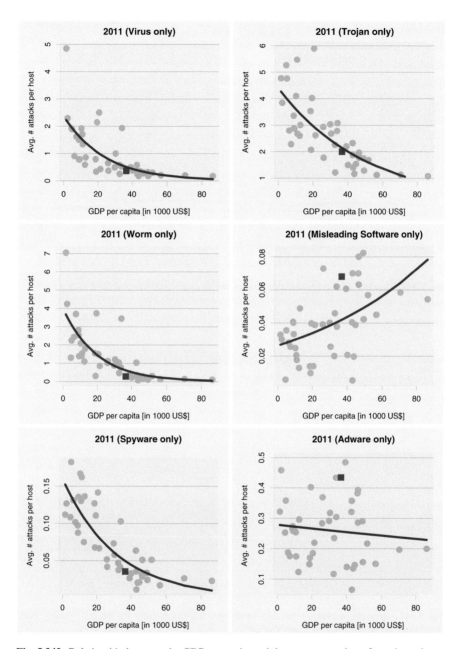

Fig. 5.342 Relationship between the GDP per capita and the average number of attacks on hosts of a country separately for virus, Trojan, worm, misleading software, spyware and adware attacks. Selected countries highlighted

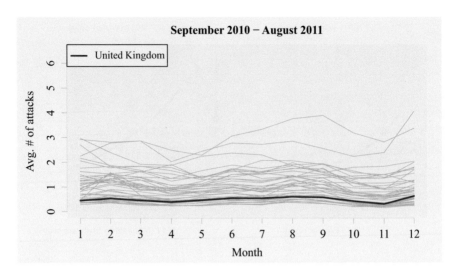

Fig. 5.343 Timeline of average monthly attacks per host by country for the 2011 time period

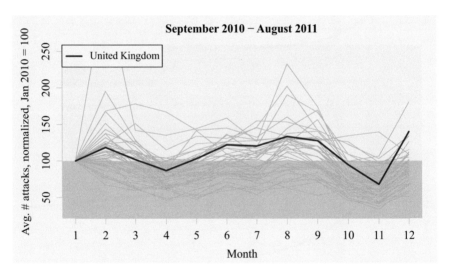

Fig. 5.344 Timeline of average monthly attacks per host by country for the 2011 time period. Normalized by attack count in September 2010

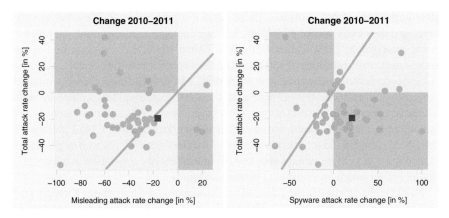

Fig. 5.345 Attack frequency changes of misleading software and spyware in relation to overall change of attack frequency

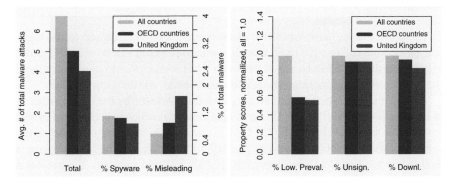

Fig. 5.346 Attack frequencies and host properties

5.44 United States

Though the US National Strategy to Secure Cyberspace [51] released by the White House in 2003 is now a dozen years old, it has been followed by a 2011 "International" Strategy for Cyberspace produced by the White House that lays out a vision for a global cybersecurity regime [50], a 2011 Department of Defense Strategy for operating in Cyberspace [52], and a 2014 document on critical infrastructure protection [53] released by the National Institute for Science and Technology. The need for a national cybersecurity strategy is therefore a clear mandate in the US with multiple diffused efforts and agencies examining this broader goal.

The 2003 White House strategy [51] has five elements: (1) design and deploy a National Cybersecurity Response System that involves a public-private partnership including a framework to assess, detect, and mitigate vulnerabilities and threats, develop cyber early warning systems, manage incident response, and share cyber threat information (2) a Vulnerability and Threat Reduction Program that better understands vulnerabilities in software and the potential resulting threats, provides law enforcement with training and mechanisms to reduce threat, secures internet protocols and provides trusted access and authentication models, comes up with advanced cybersecurity R&D efforts, (3) a broad swathe of security awareness and training programs for citizens, businesses, and government entities, (4) a framework that secures the US Government's wired and wireless networks as well as processes to manage access to those networks and (5) international cooperation in cybersecurity including watch and warning systems, and more.

In 2011, the White House released an "International Strategy for Cyberspace" [50] reflecting the White House's vision for a cyberspace that promotes freedom of expression, that promotes privacy, and that promotes the free flow of information. In order to achieve this, the strategy suggests that vendors of equipment and software intended to run on the Internet empower as wide a range of developers as possible (e.g. by providing open application program interfaces) so as to support end-to-end interoperability, and that global agreements on how to report/handle incidents and vulnerabilities be created.

The document suggests that current international laws may be enough to also regulate nation state conduct [50, p. 9] though discussions would still be needed in order to achieve a shared understanding of the issues—but continues on to suggest that these shared understandings include various economic, social, and democratic values.

A more realistic cybersecurity strategy was also disclosed by the US Department of Defense (DoD) in 2011. The US Strategic Command which is in charge of cybersecurity for the DoD created the US CyberCommand, located at Ft. Meade on the campus of the US National Security Agency with the head of the NSA wearing a dual-hat as head of US CyberCommand. For all practical purposes, this creates a single unified entity (or at least a single leader) to coordinate cybersecurity efforts between the DoD and intelligence community. A second thrust proposes new cybersecurity through improved cyber-hygiene, active defenses, and a reduction of insider threat. A third element of the DoD strategy partners the DoD with other government agencies and the Defense Industrial Basis (which is the constellation of defense contractors in the US) together with better methods to verify the trustworthiness of software and systems produced by overseas vendors. Fourth, the DoD cybersecurity strategy is linked with the White House International Cybersecurity Strategy, and aims to partner with friendly nations and traditional allies. A final element of the DoD strategy is education and training.

The US's situation w.r.t. cyber-vulnerability can be summed up in the table below.

United States	Avg number of attacks per host	Percentage of attacked hosts
2010	4.86	0.56
2011	3.51	0.49

Like most OECD countries, the US saw a significant improvement in cyberattacks from 2010 to 2011 with an over 27% decrease in the average number of attacks per host and an over 12% decrease in the percentage of attacked hosts.

The principal threat vector to the US consists overwhelmingly of Trojans, followed by viruses and worms. The number of attacks (on average) on US hosts due to misleading software and adware is much higher than one would expect from a country with the US's GDP. This suggests that cybersecurity education programs in the US should draw attention to recognizing and avoiding both misleading software and adware (Figs. 5.347, 5.348, 5.349, 5.350, 5.351, 5.352, 5.353, and 5.354).

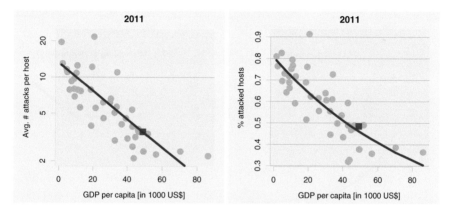

Fig. 5.347 Average number of attacks per host (*left*) and percentage of attacked host (*right*). *Blue line*: predicted values based on GDP-only model

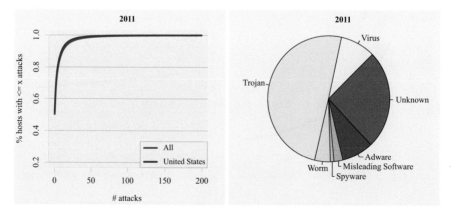

Figs. 5.348 United States: Empirical cumulative distribution of % of hosts with less than or equal to x attacks, and **5.349** United States: Distribution of attacks by type of malware

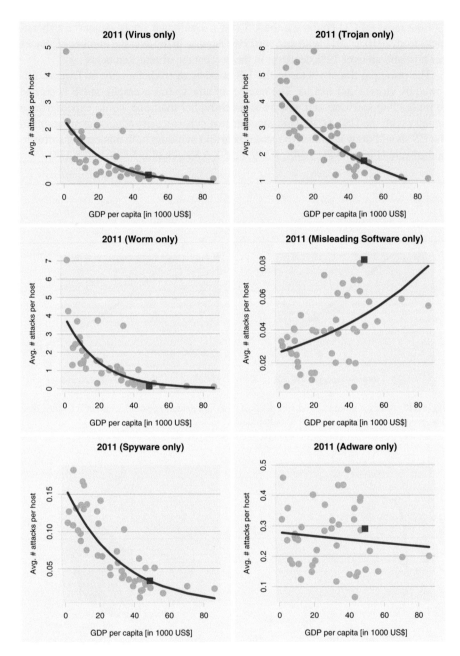

Fig. 5.350 Relationship between the GDP per capita and the average number of attacks on hosts of a country separately for virus, Trojan, worm, misleading software, spyware and adware attacks. Selected countries highlighted

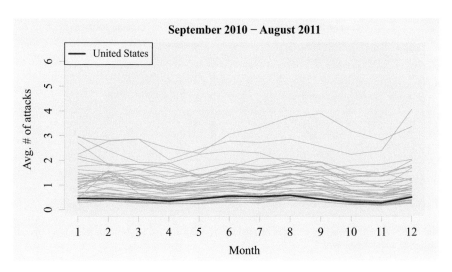

Fig. 5.351 Timeline of average monthly attacks per host by country for the 2011 time period

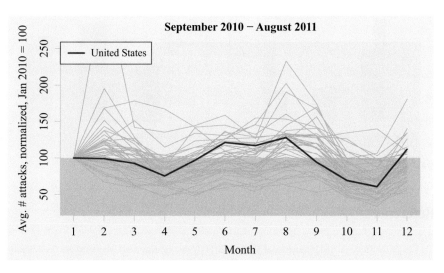

Fig. 5.352 Timeline of average monthly attacks per host by country for the 2011 time period. Normalized by attack count in September 2010

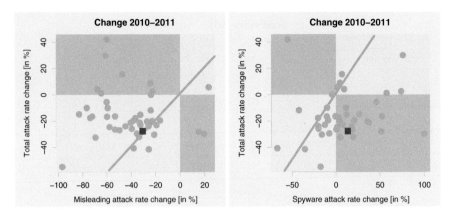

Fig. 5.353 Attack frequency changes of misleading software and spyware in relation to overall change of attack frequency

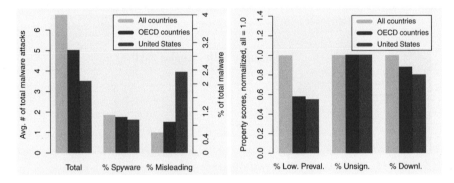

Fig. 5.354 Attack frequencies and host properties

References

1. Center for Strategic and International Studies (2011) Cybersecurity and Cyberwarfare: Preliminary Assessment of National Doctrine and Organization, http://unidir.org/files/publications/pdfs/cybersecurity-and-cyberwarfare-preliminary-assessment-of-national-doctrine-and-organization-380.pdf, *Retrieved Dec 30 2014* http://www.enisa.europa.eu/activities/Resilience-and-CIIP/national-cyber-security-strategies-ncsss/national-cyber-security-strategies-in-the-world, *Retrieved Dec 30 2014*
2. ENISA (2012). National Cyber Security Strategies Practical Guide on Development and Execution, Dec 2012
3. Wamala F (2011) ITU National Cyber Security Strategy Guide, International Telecommunication Union, Sep 2011, http://www.itu.int/ITU-D/cyb/cybersecurity/docs/ITUNationalCybersecurity StrategyGuide.pdf, *Retrieved Dec 30 2014*
4. Organization of American States (2008) A Comprehensive Inter-American Cybersecurity Strategy: A Multidimensional and Multidisciplinary Approach to Creating a Culture of

Cybersecurity, Aug 2008, http://www.oas.org/juridico/english/cyb_pry_strategy.pdf, *Retrieved Dec 30 2014*

5. Australian Government (2009) Cybersecurity Strategy, http://www.enisa.europa.eu/activities/Resilience-and-CIIP/national-cyber-security-strategies-ncsss/AGCyberSecurityStrategyforwebsite.pdf, *Retrieved Nov 26 2014*
6. Trend Micro (2008) Trend Micro Annual Threat Roundup and 2009 Forecast, 2008. *Retrieved Dec 30 2014*
7. Cyber Security Strategy .be (2012) https://www.b-ccentre.be/wp-content/uploads/2013/03/cybersecustra_fr.pdf, Nov 23 2012, Retrieved Nov 26 2014
8. Belgian Cyber Security Guide, https://www.enisa.europa.eu/activities/Resilience-and-CIIP/national-cyber-security-strategies-ncsss/belgian-cyber-security-strategy/at_download/file, *Retrieved Dec 30 2014*
9. Cyber security strategy of the Czech Republic for the 2011-2015 period, http://www.enisa.europa.eu/activities/Resilience-and-CIIP/national-cyber-security-strategies-ncsss/CzechRepublic_Cyber_Security_Strategy.pdf, Retrieved Nov 26 2014
10. Hoepers C (2005) Cybersecurity and Incident Response Initiatives: Brazil and Americas, http://www.cert.br/docs/palestras/certbr-itu-americas2005.pdf, *Retrieved Nov 26 2014*
11. Hoepers C (2012) Phishing and Banking Trojan Cases Affecting Brazil, March 2012, http://www.cert.br/docs/palestras/certbr-firstsymposium2012.pdf, *Retrieved Dec 31 2014*
12. Government of Canada (2010) Canada's Cyber Security Strategy: For a stronger and more prosperous Canada, http://www.publicsafety.gc.ca/cnt/rsrcs/pblctns/cbr-scrt-strtgy/index-eng.aspx, *Retrieved Nov 26 2014*
13. Trend Micro (2013) Latin American and Caribbean Cybersecurity Trends and Government Responses, May 2013, http://www.trendmicro.com/cloud-content/us/pdfs/security-intelligence/white-papers/wp-latin-american-and-caribbean-cybersecurity-trends-and-government-responses.pdf, *Retrieved Dec 30 2014*
14. Zhang L (2012) A Chinese Perspective on Cyber War, International Review of the Red Cross, vol 94(886):801–807 https://www.icrc.org/eng/assets/files/review/2012/irrc-886-zhang.pdf
15. Chang A (2014) Warring State: China's Cybersecurity Strategy, Center for New American Security, Dec 2014, http://www.cnas.org/sites/default/files/publications-pdf/CNAS_WarringState_Chang.pdf, *Retrieved Dec 31 2014*
16. Cyber Wellness Profile Colombia, http://www.itu.int/en/ITU-D/Cybersecurity/Documents/Country_Profiles/Colombia.pdf, Retrieved Dec 31 2014.
17. Center for Cyber Security (2015) The Danish Cyber and Information Security Strategy, Feb 2015
18. Azaria A, Richardson A, Kraus S, Subrahmanian VS (2014) Behavioral Analysis of Insider Threat: A Survey and Bootstrapped Prediction in Imbalanced Data, In: IEEE Transactions on Computational Social Systems, 1.2 (2014):135–155, Nov 2014
19. Government of Finland. Finland's Cyber Security Strategy (2013) Jan 2013. http://www.enisa.europa.eu/activities/Resilience-and-CIIP/national-cyber-security-strategies-ncsss/FinlandsCyberSecurityStrategy.pdf, Retrieved Dec 31 2014
20. Agence Nationale de La Securite des Systemes d'Information (2011) Information Systems Defense and Security: France's Strategy. http://www.enisa.europa.eu/activities/Resilience-and-CIIP/national-cyber-security-strategies-ncsss/France_Cyber_Security_Strategy.pdf, Retrieved Dec 31 2014
21. Cybersecurity Strategy for Germany (2011) Feb 2011 http://www.enisa.europa.eu/activities/Resilience-and-CIIP/national-cyber-security-strategies-ncsss/Germancybersecuritystrategy20111.pdf, Retrieved Dec 31 2014
22. Digital 21 Strategy Advisory Committee, Cybersecurity (2011) Nov 2011, http://www.digital21.gov.hk/eng/D21SAC/attachments/D21SAC_paper_9-2011.pdf, Retrieved Dec 31 2014
23. Ministry of Communication and Information Technology, National Cyber Security Policy-2013, http://www.enisa.europa.eu/activities/Resilience-and-CIIP/national-cyber-security-strategies-ncsss/NationalCyberSecurityPolicyINDIA.pdf, Retrieved Dec 31 2014.

24. Tabansky L (2013) Cyberdefense Policy of Israel: Evolving Threats and Responses, Chair de Cyberdefense et Cybersecurite, Jan 2013—Article n° III.12
25. Tabansky L, Aviv I (2013) "Critical National Infrastructure Protection: Evolution of Israeli Policy." In: Proceedings of the 12th European Conference on Information Warfare and Security: ECIW 2013, Academic Conferences Limited
26. Presidency of the Council of Ministers (2013) National Strategic Framework for Cyberspace Security, Dec 2013 http://www.enisa.europa.eu/activities/Resilience-and-CIIP/national-cyber-security-strategies-ncsss/IT_NCSS.pdf, Retrieved Dec 31 2014
27. Information Security Policy Council, Cybersecurity Strategy, June 2013. http://www.enisa.europa.eu/activities/Resilience-and-CIIP/national-cyber-security-strategies-ncsss/JAP_NCSS2.pdf, Retrieved Dec 31 2014
28. Guri, M., Kedma, G., Kachlon, A., & Elovici, Y. (2014, October). AirHopper: Bridging the air-gap between isolated networks and mobile phones using radio frequencies. In *Malicious and Unwanted Software: The Americas (MALWARE), 2014 9th International Conference on* (pp. 58–67). IEEE.
29. Hashim MSB (2009) Malaysia's National Cyber Security Policy, http://www.itu.int/ITU-D/cyb/events/2009/tunis/docs/hashim-cybersecurity-malaysia-june-09.pdf, Retrieved Nov 26 2014
30. Ministry of Science, Technology, and Innovation (Malaysia) (2008) National Cyber Security
31. Trend Micro and Organization of American States (2015) Report on Cybersecurity and Critical Infrastructure in the Americas, Retrieved June 15 2015.
32. National Coordinator for Security and Counterterrorism, National Cyber Security Strategy 2, 2013,http://www.enisa.europa.eu/activities/Resilience-and-CIIP/national-cyber-security-strategies-ncsss/NCSS2Engelseversie.pdf, Retrieved Dec 31 2014
33. Ministry of Defense (Netherlands), The Defense Cyber Strategy, 2012 https://ccdcoe.org/strategies/Defence_Cyber_Strategy_NDL.pdf, Retrieved Dec 31 2014
34. New Zealand Government, New Zealand's Cyber Security Strategy, 2011, http://www.enisa.europa.eu/activities/Resilience-and-CIIP/national-cyber-security-strategies-ncsss/nzcybersecuritystrategyjune2011_0.pdf, Retrieved Dec 31 2014
35. The Ministry of Government Administration, Reform and Church Affairs (Norway), Cyber Security Strategy for Norway, https://www.regjeringen.no/globalassets/upload/FAD/Vedlegg/IKT-politikk/Cyber_Security_Strategy_Norway.pdf, Retrieved June 15 2015
36. National Cybersecurity Coordination Office, Philippine Cyber Security Efforts, Aug 2010. https://www.itu.int/ITU-D/asp/CMS/Events/2010/NGN-Philippines/S5-Philippines_cybersecurity.pdf, Retrieved Dec 31 2014
37. Philippine National Cyber Security Plan (2005) http://phcybersecurity.blogspot.com/2011/09/philippine-national-cyber-security-plan.html, Retrieved June 15 2015
38. Republic of Poland, Ministry of Administration & Digitization, Internal Security Agency, Cyber-Space Protection Policy of the Republic of Poland, June 2013. https://www.enisa.europa.eu/activities/Resilience-and-CIIP/national-cyber-security-strategies-ncsss/copy_of_PO_NCSS.pdf, Retrieved Dec 31 2014.
39. Information Security Doctrine of the Russian Federation, Sep 2010, http://archive.mid.ru//bdomp/ns-osndoc.nsf/1e5f0de28fe77fdcc32575d900298676/2deaa9ee15ddd24bc32575d9002c442b!OpenDocument, Retrieved Dec 31 2014
40. Ministry of Communication and Information Technology (Saudi Arabia), Developing National Information Security Strategy for the Kingdom of Saudi Arabia, http://www.mcit.gov.sa/Ar/MediaCenter/PubReqDocuments/NISS_Draft_7_EN.pdf, Retrieved June 15 2015
41. Saudi Aramco Cyber Attacks a 'wake-up call', Says Former NSA Boss, InfoSecurity, May 8 2014, http://www.infosecurity-magazine.com/news/saudi-aramco-cyber-attacks-a-wake-up-call-says/ Retrieved June 15 2015
42. Infocomm Development Authority of Singapore, National Cyber Security Masterplan 2018, http://www.ida.gov.sg/~/media/Files/Programmes%20and%20Partnership/Initiatives/2014/ncsm2018/NationalCyberSecurityMasterplan%202018.pdf, Retrieved July 20 2015

43. Infocomm Development Authority of Singapore, Annex A: Factsheet on National Cyber Security Masterplan 2018, https://www.ida.gov.sg/~/media/Files/About%20Us/Newsroom/Media%20Releases/2013/0724_ncsm/AnnexA.pdf, Retrieved July 10 2015
44. Department of Communications, Notice of Intention to Make South African National Cybersecurity Policy, Feb 2010 http://www.enisa.europa.eu/activities/Resilience-and-CIIP/national-cyber-security-strategies-ncsss/southafricancss.pdf, Retrieved Dec 31 2014
45. National Cyber Security Masterplan (South Korea), Aug 2011 https://ccdcoe.org/sites/default/files/strategy/KOR_NCSS_2011.pdf, Retrieved Dec 31 2014
46. National Cyber Security Strategy (Spain) (2013) http://www.enisa.europa.eu/activities/Resilience-and-CIIP/national-cyber-security-strategies-ncsss/NCSS_ESen.pdf, Retrieved Dec 31 2014
47. Federal Department of Defence, Civil Protection and Sport DDPS (Switzerland), National strategy for the protection of Switzerland against cyber risks, June 2012 http://www.enisa.europa.eu/activities/Resilience-and-CIIP/national-cyber-security-strategies-ncsss/National_strategy_for_the_protection_of_Switzerland_against_cyber_risksEN.pdf, Retrieved Dec 31 2014
48. Ministry of Transport, Maritime Affairs, and Communications (Turkey), National Cyber Security Strategy and 2013-2014 Action Plan, 2013 http://www.enisa.europa.eu/activities/Resilience-and-CIIP/national-cyber-security-strategies-ncsss/TUR_NCSS.pdf, Retrieved Dec 31 2014
49. Cabinet Office (UK), The UK Cyber Security Strategy, 2011 http://www.enisa.europa.eu/activities/Resilience-and-CIIP/national-cyber-security-strategies-ncsss/UK_NCSS.pdf, Retrieved Dec 31 2014
50. The White House, International Strategy for Cyber Space, May 2011 http://www.enisa.europa.eu/activities/Resilience-and-CIIP/national-cyber-security-strategies-ncsss/international_strategy_for_cyberspace_US.pdf, Retrieved Dec 31 2014
51. The White House, The National Strategy to Secure Cyber Space, Feb 2003 https://www.us-cert.gov/sites/default/files/publications/cyberspace_strategy.pdf, Retrieved Dec 31 2014
52. Department of Defense (USA), Department of Defense Strategy for Operating in Cyber Space, July 2011 http://www.defense.gov/news/d20110714cyber.pdf, Retrieved Dec 31 2014
53. National Institutes of Standards and Technology, Framework for Improving Critical Infrastructure Cybersecurity, Feb 2014 http://www.nist.gov/cyberframework/upload/cybersecurity-framework-021214-final.pdf, Retrieved June 15 2015
54. Austrian Cyber Security Strategy (2013) http://www.enisa.europa.eu/activities/Resilience-and-CIIP/national-cyber-security-strategies-ncsss/AT_NCSS.pdf, Retrieved Nov 26 2014
55. Presidencia da Republica (Brazil) (2012) Overview of Information Security (Cyber Security and Cyber Defense) of Critical Infrastructure in Brazil, Sep 2012, http://www.oas.org/cyber/presentations/OAS%20set%202012.pdf, *Retrieved Nov 26, 2014*

Index

© Springer International Publishing Switzerland 2015
V.S. Subrahmanian et al., *The Global Cyber-Vulnerability Report*, Terrorism,
Security, and Computation, DOI 10.1007/978-3-319-25760-0